Andreas Gubler

Quantitative Estimations of Soil Properties by VNIR Spectroscopy

Andreas Gubler

Quantitative Estimations of Soil Properties by VNIR Spectroscopy

Applications for Laboratory and Field Measurements

Südwestdeutscher Verlag für Hochschulschriften

Impressum/Imprint (nur für Deutschland/only for Germany)
Bibliografische Information der Deutschen Nationalbibliothek: Die Deutsche Nationalbibliothek verzeichnet diese Publikation in der Deutschen Nationalbibliografie; detaillierte bibliografische Daten sind im Internet über http://dnb.d-nb.de abrufbar.
Alle in diesem Buch genannten Marken und Produktnamen unterliegen warenzeichen-, marken- oder patentrechtlichem Schutz bzw. sind Warenzeichen oder eingetragene Warenzeichen der jeweiligen Inhaber. Die Wiedergabe von Marken, Produktnamen, Gebrauchsnamen, Handelsnamen, Warenbezeichnungen u.s.w. in diesem Werk berechtigt auch ohne besondere Kennzeichnung nicht zu der Annahme, dass solche Namen im Sinne der Warenzeichen- und Markenschutzgesetzgebung als frei zu betrachten wären und daher von jedermann benutzt werden dürften.

Coverbild: www.ingimage.com

Verlag: Südwestdeutscher Verlag für Hochschulschriften GmbH & Co. KG
Heinrich-Böcking-Str. 6-8, 66121 Saarbrücken, Deutschland
Telefon +49 681 37 20 271-1, Telefax +49 681 37 20 271-0
Email: info@svh-verlag.de

Approved by: Bern, Universität, Diss., 2011

Herstellung in Deutschland (siehe letzte Seite)
ISBN: 978-3-8381-3361-4

Imprint (only for USA, GB)
Bibliographic information published by the Deutsche Nationalbibliothek: The Deutsche Nationalbibliothek lists this publication in the Deutsche Nationalbibliografie; detailed bibliographic data are available in the Internet at http://dnb.d-nb.de.
Any brand names and product names mentioned in this book are subject to trademark, brand or patent protection and are trademarks or registered trademarks of their respective holders. The use of brand names, product names, common names, trade names, product descriptions etc. even without a particular marking in this works is in no way to be construed to mean that such names may be regarded as unrestricted in respect of trademark and brand protection legislation and could thus be used by anyone.

Cover image: www.ingimage.com

Publisher: Südwestdeutscher Verlag für Hochschulschriften GmbH & Co. KG
Heinrich-Böcking-Str. 6-8, 66121 Saarbrücken, Germany
Phone +49 681 37 20 271-1, Fax +49 681 37 20 271-0
Email: info@svh-verlag.de

Printed in the U.S.A.
Printed in the U.K. by (see last page)
ISBN: 978-3-8381-3361-4

Copyright © 2012 by the author and Südwestdeutscher Verlag für Hochschulschriften GmbH & Co. KG and licensors
All rights reserved. Saarbrücken 2012

Summary

Soils are a limited resource. Information about soils is needed to assess and monitor its quality and to ensure a sustainable soil management. The acquisition of soil data is limited by the available amounts of time and money. Visible and Near Infrared Spectroscopy (VNIRS) measures the reflectance of wavelengths between 300 and 2500 nm and represents a novel, non-destructive technique allowing for fast and inexpensive soil analyses. Thus, it is considered as valuable alternative to retrieve soil data more efficiently.

In the present work, models to estimate organic carbon (C_{org}) and total nitrogen (N_{tot}) contents from spectra of dried and sieved soil samples were derived for different sets of soil samples covering local to regional to national (mostly Swiss) areas. The assessed calibration algorithms included partial least squares regression (PLSR) and wavelet transforms in combination with quadratic regression models. Furthermore, soil samples were rewetted under laboratory conditions to assess the influence of moisture on soil reflectance. Different parameters derived from the spectra were tested for their suitability to estimate soil water contents. And finally, a portable VNIRS device (FieldSpec 3) was mounted to a mobile platform to conduct VNIRS field measurements. Two transects were measured by stop-and-go as well as on-the-go at the long-term fertilisation experiment Bad Lauchstädt (Saxony-Anhalt, Germany) to estimate C_{org} and N_{tot}.

For datasets of local to regional extensions, root mean squared errors (RMSE) comparable to published studies were achieved for C_{org} and N_{tot} estimates from spectra of dried and sieved soil samples (2 to 3 g C kg^{-1} and 0.2 to 0.3 g N kg^{-1}, respectively). For sets of national extension, higher RMSE than expected were observed, presumably due to the relatively small numbers of soil samples with respect to the soil variability included. C_{org} estimates of validation samples taken from another ensemble of soil samples than those used for calibration exhibited good correlation with the reference analyses, but were biased. The accuracy as well as the reproducibility of the models' estimates strongly depended on the number of PLS factors or wavelet coefficients, respectively, included in the models.

The reflectance over the whole VNIR range was effected by soil moisture, but certain wavelengths exhibited an augmented sensitivity, especially for low water contents. A parameter capturing the relative extension of the water absorption band near 1940 nm was found most useful to estimate soil water content. The field measurements by the mobile platform were strongly influenced by soil moisture. While relative differences of C_{org} and N_{tot} were visible from on-the-go field measurements even when using a calibration derived from dried samples, specific calibrations were needed to retrieve absolute levels. Reference analyses of nine samples sufficed to adapt the existing calibration.

It was concluded that VNIRS measurements from dried soil samples could become a equivalent alternative to traditional analytical methods given that measuring procedures will be standardised and internal standards to compare spectra of different spectrometers will be available. On-the-go field measurements were considered useful for small-scale applications like precision farming, whereas for the larger scales required for soil mapping it was recommended to collect soil samples for laboratory measurements.

Acknowledgements

Thousand thanks to the many persons who contributed in some way or another to the present work. Credit goes to:

Professor Peter German for enabling the present work within his group, for discussions and suggestions.

Marco Carizzoni for his generous support in many areas during the last three and a half years.

Bernard Barthès, IRD Montpellier, the second examiner of this work, for his efforts concerning the manuscript presented in chapter 4 and the fruitful and motivating meeting in Vienna.

Isabel Richli, Benjamin Huber, Dino Andrini, and Matthias Wiggenhauser for their work and commitment during their Bachelor theses which all contributed considerably to this work.

All institutions that provided soil samples and/or analytical data: Reto Meuli, Stefan Amman, and Peter Schwab, NABO/ART Reckenholz; Stéphane Burgos, EIC Nyon; Franz Borer and Gaby von Rohr, Fachstelle Bodenschutz Kanton Solothurn; Nicolas Rossier, FRIBO; Stephan Zimmermann and Peter Lüscher, WSL Birmensdorf; Janine Krüger, Uwe Franko, Ines Merbach, Claudia Dierke, and Ulrike Werban, UFZ Leipzig/Halle; Christine Hauert and Lorenz Ruth, CDE University Bern.

Dagmar Hensel and Ingrid Hincapié – Jürg Schenk for technical support and uncounted hours of floorball – Daniela Fischer for laboratory advise – Library team of the Institute of Geography – Basil Ferrante – Hans Hurni, Urs Balsiger, and the rest of CDE – RSL University of Zürich – all partners of the iSOIL project – the colleagues from the soil spectroscopy community for good times at conferences and workshops – everyone I forgot ...

And last but certainly not least, I thank Sylvia and Anna for their patience and understanding, especially during the last three months.

This work was financed by the FP7 collaborative project iSOIL – Interactions between soil related sciences - Linking geophysics, soil science and digital soil mapping, no. 211386.

Contents

List of Tables   v

List of Figures   vii

List of Abbreviations and Symbols   xi

1. Introduction   1
    1.1. Visible and Near Infrared Spectroscopy (VNIRS) 2
    1.2. VNIRS applications 2
    1.3. The present work 3

2. Methods   5
    2.1. Principles of Visible and Near Infrared Spectroscopy (VNIRS) 5
    2.2. Soil samples and archives 9
    2.3. VNIRS laboratory measurements 10
    2.4. Reference analyses 12
    2.5. Calibration 13
    2.6. Validation statistics 23
    2.7. Software: Using R 25

3. VNIRS laboratory models   27
    3.1. Estimating soil carbon 27
    3.2. Estimating total nitrogen 33
    3.3. Applying existing calibrations 37
    3.4. Reproducibility of VNIRS measurements 41
    3.5. Summary and Conclusions 46

4. Comparing different calibration algorithms using wavelets   49
    4.1. Materials and methods 50
    4.2. Results 53
    4.3. Discussion 55
    4.4. Summary and Conclusions 58

5. Soil moisture effects on VNIR reflectance   61
    5.1. Water absorption features 61
    5.2. Methods 63
    5.3. Results 67
    5.4. Discussion 69
    5.5. Summary and Conclusions 73

6. Field measurements — **75**
- 6.1. State of the art 76
- 6.2. Methods 79
- 6.3. Results 83
- 6.4. Discussion 89
- 6.5. Summary and Conclusions 95

7. Conclusions and perspectives — **97**
- 7.1. Laboratory applications 97
- 7.2. Field applications 98
- 7.3. The need for standardisation 98
- 7.4. Model stability 99
- 7.5. Soil mapping 100

A. Additional information — **103**
- A.1. Chapter 2: Methods 103
- A.2. Chapter 3: VNIRS laboratory models 112
- A.3. Chapter 5: Soil moisture effects 119

Bibliography — **135**

List of Tables

2.1.	Characteristics of the visible, the near infrared (NIR), and the mid infrared (MIR) ranges of electromagnetic radiation.	6
2.2.	Important VNIR absorption bands observed for soil samples	8
3.1.	Studied sets of soil samples to estimate carbon contents	28
3.2.	Performance of PLSR models (cross-validation RMSE) to estimate carbon.	30
3.3.	Studied sets of soil samples to estimate nitrogen contents	34
3.4.	Performance of PLSR models (cross-validation RMSE) to estimate N_{tot}.	35
4.1.	Studied sets of soil samples for wavelet models	51
5.1.	Assessed indicators for water content.	65
5.2.	Performance of different models to estimate soil water content ω by parameters deduced from reflectance data and the auxiliary variable C_{tot}.	69
5.3.	Soil samples used for figure 5.5.	70
6.1.	Specifications of the mobile platform constructed for the present work.	79
6.2.	Performance of VNIRS calibrations for Bad Lauchstädt depending on the spectral pre-treatments	86
A.1.	Soil samples used for assessment of soil moisture effects on spectroscopic measurements.	119

List of Figures

2.1.	Schematic of specular and diffuse reflection.	5
2.2.	Reflectance spectra of five randomly selected soil samples in oven dry condition.	7
2.3.	Used laboratory measuring setup: FieldSpec 3 spectrometer, muglight, and petri dish containing soil material placed atop of the muglight for the spectral measurement.	10
2.4.	Comparison of C_{org} contents determined by dry combustion (CN analyser) and dichromate digestion according to the Swiss standard method.	13
2.5.	RMSE as a function of the number of PLS factors used for the 'fr-so' model to estimate organic matter presented in section 3.1	17
2.6.	SD-OD plot and leverage plot for the 'fr-so' model to estimate organic matter presented in section 3.1	19
2.7.	The Haar wavelet and Daubechies' so-called extremal phase wavelet with eight vanishing moments.	20
2.8.	Haar wavelets for the scales 0 to 2.	21
2.9.	Illustration of Mallat's pyramid algorithm for the discrete wavelet transform for a data vector of length $16 = 2^4$.	21
3.1.	PLSR models to estimate organic and total carbon: RMSE and model estimates.	29
3.2.	RMSE and standard deviation of models to estimate total and organic carbon compared to those of published works.	31
3.3.	Regression coefficients of the derived PLSR models to estimate organic and total carbon.	32
3.4.	Scatter plot of the first three principle components of the pooled dataset of all absorbance spectra used for the different PLSR models to estimate total and organic carbon.	33
3.5.	PLSR models to estimate total nitrogen: RMSE and model estimates.	35
3.6.	Regression coefficients of the derived PLSR models to estimate total nitrogen.	36
3.7.	Score and orthogonal distances and OM estimates of the *lu* samples by the *fr-so* model including 13 PLS factors	38
3.8.	RMSE, bias, and variance in respect of the included number of PLS factors for C_{org} estimates of the *lu* samples by *fr-so* models.	39
3.9.	Standard error of laboratory (left) and bias of C_{org} estimates for duplicate analyses by the same VNIRS device as well as by different devices.	42
3.10.	Reflectance spectra of fine sand as recorded by two different VNIRS devices.	43
3.11.	Reproducibility of C_{org} analyses by VNIRS, dichromate digestion, and dry combustion.	44
4.1.	Example of wavelet decomposition for one randomly selected soil sample	50
4.2.	Comparison of calibration RMSE for N_{tot} estimates by algorithms using variance, covariance, and correlation coefficients as selection criterion for wavelet coefficients	53

4.3. Comparison of calibration RMSE for N_{tot} estimates by algorithms using variance and median absolute deviation as selection criterion for wavelet coefficients, and with PLSR. 54
4.4. Comparison of calibration RMSE for OM estimates by algorithms using variance, covariance, and correlation coefficients as selection criterion for wavelet coefficients 54
4.5. Comparison of calibration RMSE for OM estimates by algorithms using variance and median absolute deviation as selection criterion for wavelet coefficients, and with PLSR. 54
4.6. Comparison of validation RMSE for OM estimates by algorithms using variance, covariance, and correlation coefficients as selection criterion for wavelet coefficients 56
4.7. Comparison of validation RMSE for OM estimates by algorithms using variance and median absolute deviation as selection criterion for wavelet coefficients, and with PLSR. 56
4.8. Scale and position of the first 20 wavelet coefficients selected by the algorithms using variance, covariance, correlation coefficients, and MAD for N_{tot} and OM estimations 57

5.1. NIR spectrum of solid water and the most important absorption features 61
5.2. Changes in reflectance of sand during drying out. 62
5.3. Flow chart: Measuring procedure for moist (and air dry) soil samples. 64
5.4. Reflectance spectra of sample nine at different soil water contents ω, and illustration of the parameters deduced from the spectra for data analysis 66
5.5. Influence of soil moisture on some of the selected indicators deduced from spectral data for seven randomly selected soil samples. 68
5.6. Correlation between water content ω, F_1, and reflectance ratio 1940:1840 nm for all soil samples. 70

6.1. Shank-based spectrophotometer used by Christy (2008) to obtain NIR spectra . . 78
6.2. Schematic illustration of the NIR spectrophotometer sensor used for on-the-go measurements presented by Mouazen et al. (2005a, illustration therein) 78
6.3. Setup used by the University of Bern for VNIRS field measurements by an ASD FieldSpec 3 80
6.4. Long-term fertilisation experiment Bad Lauchstädt, plot seven: combinations of N, P, and K fertilisers and levels of manure applied to subplots 1 to 18. 80
6.5. Top soil concentrations of C_{org} and N_{tot} as well as C/N ratios determined by dry combustion for plot seven. 81
6.6. Reflectance spectra collected at transect one from subplot two by different measuring setups 84
6.7. Cross-validation and validation N_{tot} estimates by VNIRS for calibrations based on the different spectral pre-treatments. 85
6.8. Comparison of C_{org} and N_{tot} contents of transect one, plot seven, determined by VNIRS laboratory measurements and CN analyser. 87
6.9. Comparison of C_{org} and N_{tot} VNIRS estimates from field (in situ) stop-and-go, on-the-go, and laboratory measurements of transect one; gravimetric water content determined for laboratory samples. 88

6.10. C_{org} and N_{tot} estimates for VNIRS field measurements of transect one using calibration models based on laboratory spectra spiked with field spectral data, spiked with laboratory and field spectral data, and using calibrations based on field measurements only of nine local validation samples 90
6.11. C_{org} and N_{tot} estimates for VNIRS field measurements of transect two using calibration models based on laboratory spectra spiked with field spectral data, spiked with laboratory and field spectral data, and using calibrations based on field measurements only of nine local validation samples 91
6.12. Ratio and difference of field and laboratory VNIRS estimates for C_{org} versus the gravimetric water content of corresponding soil samples of transect one. 94

7.1. Flow chart: Proposed procedure to implement VNIRS in soil mapping. 101

A.1. Elements used for flow charts. 103
A.2. Flow chart: Procedure for laboratory VNIRS measurements. 104
A.3. Flow chart: Pre-treatments of spectral data. 105
A.4. Flow chart: Calibration by partial least squares regression (PLSR) for target variable y. 106
A.5. Flow chart: Estimation of unknown samples by PLSR. 107
A.6. Flow chart: Calibration for target variable y by models based on wavelet coefficients. 108
A.7. Flow chart: Estimation of unknown samples by a model based on wavelet coefficients. 109
A.8. Flow chart: Principle of cross-validation for regression models. 110
A.9. Scatter plots of the first five principle components for the pooled dataset of all absorbance spectra used for the different PLSR models to estimate total and organic carbon. 112
A.10. Absorbance spectra of five randomly selected soil samples in oven dry condition. 113
A.11. Correlation of C_{org} and N_{tot} for the *rosslau* and *lu* datasets. 113
A.12. Score distances and orthogonal distances of *lu* samples in respect of C_{org} models derived from the *fr-so* dataset. 114
A.13. Scatter plots of C_{org} estimates for *lu* samples by models derived from the *fr-so* dataset. 115
A.14. First and second C_{org} analyses of *lu* samples by models derived from the *fr-so* dataset. 116
A.15. Reflectance spectra of fine sand recorded for different measurement series. 117
A.16. Reflectance ratio of FS Pro to FS3 spectra of the sand sample used as standard. . 117
A.17. Reflectance of the Spectralon panel used for white reference relative to an unused spectralon panel for different dates. 117
A.18. Comparison of 20 measurements of a Spectralon panel with and without dislocation of the panel between measurements. 118
A.19. Reflectance spectra and reflectance ratio wet:dry of all soil samples for different water contents 120
A.20. Influence of soil moisture on reflectance at eight different wavelengths for seven randomly selected soil samples. 129
A.21. Influence of soil moisture on reflectance ratios for seven randomly selected soil samples. 131

List of Abbreviations and Symbols

A	Absorbance
a	Number of PLS factors
ASD	Analytical Spectral Devices Inc., Boulder, CO, USA (manufacturer of VNIR spectrometers)
CV	Coefficient of variation (equation 2.23)
C_{org}	Organic carbon (g kg^{-1})
C_{tot}	Total carbon (g kg^{-1})
F_1, F_2	Relative areas; parameters derived from spectral data (equations 5.3, 5.4)
MAD	Median absolute deviation (robust measure for variability; equation 4.1)
MD	Mahalanobis distance (equation 2.7)
MSE	Mean squared error (equation 2.21)
MIR	Mid infrared (roughly from 2500 to 25 000 nm wavelengths)
NIR	Near infrared (roughly from 760 to 2500 nm wavelengths)
N_{tot}	Total nitrogen (g kg^{-1})
OD	Orthogonal distance (equation 2.17)
OM	Organic matter
PCA	Principle component analysis
PLS(R)	Partial least squares (regression)
R^2	Coefficient of determination (equation 2.28)
RMSE	Root mean squared error (equation 2.22)
r_λ	Reflectance ratio 1940:1840 nm wavelengths
SD	Score distance (equation 2.9)
sd	Standard deviation
SEL	Standard error of laboratory (equation 2.26)
T	Score matrix $(t_1^T, t_2^T, \ldots, t_a^T)$
t_i	Score vector
$\tilde{v}$	Wavenumber (cm^{-1})
VNIR(S)	Visible and near infrared (spectroscopy)
χ^2	Quantile of the chi-squared distribution
λ	Wavelength (nm)
ω	Gravimetric soil water content (kg kg^{-1})
$\Box^T$	Matrix transposition
$\overline{\Box}$	Mean of $\Box$
$\hat{y}$	Estimate of y

1. Introduction

> Buy land, they're not making it anymore.
> Mark Twain (1835-1910)

Although scientists disagree that land is not produced anymore, Twain's ironic statement reflects that land – and particularly arable land – is a limited resource. Land is required by numerous groups and individuals for very different activities and purposes: it is needed for agricultural and forestal production of food and raw materials, it is consumed to build villages, cities, industrial production plants, and all the infrastructures required for our style of living, it is demanded as space for leisure and recreation, it is abused to dispose any type of waste, and some people fight to preserve some areas as natural biotopes. Therefore, conflicts are inevitable, especially in densely populated areas. In Switzerland for example, the settlement and infrastructure area increased by 23 % or 490 km^2 between 1983 and 2007, and most of the converted land was of agricultural origin (FSO, 2011).[1]

Land-use planning and regulation aim to ensure an efficient, fair, and sustainable use of land and soils (Hepperle & Lendi, 1993). This duty is complicated by the fact that soil characteristics are differing strongly due to differences in pedological origin, climatic and topographic conditions, and anthropogenic influences. Sustainable use of land and soils categorically implies the preservation of good soils, because soils are rapidly degrading if managed inadequately, but are not emerging or recovering within time spans relevant to humans (Gisi et al., 1997). To achieve the commitments related to land-use planning and soil monitoring, the involved authorities and scientists depend on sound and reliable information about soils and their characteristics, e.g. incorporated into soil maps. The available data are limited by the spendable amounts of time and money. In addition to traditional soil mapping and classical chemical analyses, supplementary approaches like geostatistical methods, remote sensing, and novel, fast and inexpensive analysis techniques could allow to use the available resources more efficiently. The data produced by these new approaches are possibly slightly less accurate than those by classical ones, but this disadvantage is outbalanced by the large quantities of data produced. Amongst others, visible and near infrared spectroscopy (VNIRS) is considered as promising technique to generate soil data at reasonable expenses (McBratney et al., 2006).

Beside for planning and monitoring, large amounts of soil data are for example needed in precision farming. This agricultural technique regards small-scale variations within the cultivated fields to adapt the inputs of fertiliser, herbicides, and other resources. As a result, agricultural goods are produced more efficiently and with less ecological impacts (Maleki et al., 2008). VNIRS is a promising technique in this field because it can be used with an on-the-go setup. In addition, VNIRS is also an interesting tool to assess scientific problems, e.g. the question whether soils act as carbon sinks or sources in the context of global warming. There is evidence that carbon storage

---

[1] Total area of Switzerland: 41 300 km^2 whereof 26 % non-productive areas (water, rock, glaciers, ...). In 1997, settlement and urban areas made up 9 % (or 2 800 km^2) of the productive areas; agricultural and wooded areas accounted for 50 and 41 %, respectively (FSO, 2001).

of soils can be promoted by appropriate soil management, but the uncertainty in understanding the causes, magnitude and permanence of soil carbon sinks is still substantial (Read et al., 2001). Usually, large numbers of soil analyses are required to detect the rather small changes in soil organic carbon due to its spatial and seasonal variability. Of course, there are numerous other scientific questions where data generated by VNIRS could facilitate a serious examination of the problem.

1.1. Visible and Near Infrared Spectroscopy (VNIRS)

Visible and Near Infrared Spectroscopy (VNIRS) – sometimes referred to as VisNIR spectroscopy or diffuse reflectance spectroscopy (DRS) – is considered as adequate technology to collect efficiently information about soils for various purposes like soil monitoring, soil mapping, and precision agriculture (McBratney et al., 2003, 2006; Viscarra Rossel et al., 2006b; Stenberg et al., 2010). In other areas like food technology and agriculture, VNIRS has been established as analytical technique for decades as a result of the work by K.H. Norris and co-workers (e.g. Ben-Gera & Norris, 1968). Although first investigations on VNIR soil reflectance were published over 45 years ago (Bowers & Hanks, 1965), there has been a growing interest and a lot of investigation in introducing this technique for soil analysis for the last 20 years only. The development of small, portable devices enhanced the interest in VNIRS resulting in a surge of published studies. VNIRS devices allow measuring the diffuse reflectance in the range of visible light (400-760 nm wavelengths) and the near infrared (NIR, 760-2500 nm). This portion of the electromagnetic spectrum is dominated by overtones and combination bands of C-H, O-H, and N-H functionalities (Workman & Weyer, 2007). VNIR spectra are rich on information, but the information is confounded because absorption bands are weak and overlapping and optical characteristics between single samples can change (Workman & Shenk, 2004). Therefore, statistical methods like partial least squares regression (PLSR) are required to derive models from the collected spectral data.

1.2. VNIRS applications

Different VNIRS applications were proposed for soil analyses. Roughly, they can be summarised in three groups:

Laboratory spectroscopy Properties of soil samples are analysed by measuring the diffuse reflectance of dried, sieved, and possibly milled samples in the laboratory. Laboratory spectroscopy is considered as alternative to traditional analytical techniques allowing fast and inexpensive analyses (Viscarra Rossel et al., 2006b).

Proximal sensing/in-situ measurements Portable VNIRS devices allow measurements directly in the field. Using a contact probe (which contains an illumination source and the spectrometer's optical sensor), only little preparation is required to measure soil layers within profiles or intact soil cores (Viscarra Rossel et al., 2009). Alternatively, the soil surface's reflectance is measured from a short distance using sun light or an artificial illumination. **On-the-go sensors** may be considered as special case of proximal sensing: the reflectance of whole fields and transects is measured by sensors mounted to a mobile platform. The sensors may be invasive or not. Depending on the design, the soil surface or a portion of the top layer is captured (Christy, 2008; Mouazen et al., 2005a).

Remote sensing The diffuse reflectance of a field or an entire region is analysed by airborne devices or satellites. As some of the atmosphere's constituents – primarily H_2O and CO_2 – exhibit strong absorptions in the NIR, the range of suitable wavelengths is reduced for these measurements. The incident radiation is almost completely removed between 1350 and 1500 nm as well as between 1800 and 2000 nm wavelengths. In the field of soil analyses, airborne and satellite VNIR spectroscopy could not meet the big expectations so far (e.g. Stevens et al., 2006, 2008). In addition, the use is limited because remote VNIRS only captures the surface, thus wherever vegetation is present, the spectral data refer to the vegetation and not the soil.

Numerous soil parameters were investigated by VNIRS during the last years: soil moisture, texture (most often the clay fraction), a huge variety of chemical elements from Al to Zn as well as soil biological properties like biomass and respiration rate (Viscarra Rossel et al., 2006b; Stenberg et al., 2010). Most often, VNIRS was used to estimate total and organic carbon and its fractions, and usually good results were achieved. Clay content, iron oxides, and cation exchange capacities are other properties targeted often and successfully by VNIRS.

Beside from its use as analytical technique, VNIRS has a tremendous potential which has only been scarcely used so far. Soil spectral data are a holistic approach to soil because many physical, chemical, and biological parameters contribute to soil reflectance. Therefore, the assessment of soil variability and changes in soil conditions based directly on spectral data could become an important tool for soil mapping and soil monitoring (Odlare et al., 2005; Islam et al., 2005; Demattê et al., 2004). The intermediary step of estimating soil properties could be skipped. If chemical analyses still are desired, the information about spectral soil variability can be used to derive an efficient sampling scheme. Yet another possibility was proposed by McBratney et al. (2006) who suggested to estimate pedotransfer functions directly by spectral data. The described extensions of soil spectroscopy are beyond the scope of the present dissertation, but I believe that these approaches will be very fruitful.

1.3. The present work

The present work was conducted within the framework of the European collaborative project *iSOIL: Interactions between soil related sciences – Linking geophysics, soil science and digital soil mapping*[2] which focused on 'improving fast and reliable mapping of soil properties, soil functions and soil degradation threats' (iSOIL, 2008). The project aimed to improve and integrate geophysical and spectroscopic measuring techniques and combine them with advanced approaches for soil sampling and pedometrics. Within the iSOIL project, VNIR spectroscopy was considered as emerging technique that should be integrated on mobile platforms together with other devices like electromagnetic induction (EMI) mappers, ground penetrating radar (GPR) and γ-ray spectrometers.

For this purpose, different soil archives were captured by VNIRS to derive calibrations to estimate soil properties from spectral data (chapter 3). Moreover, novel calibration algorithms

---

[2] Grant Agreement number 211386; co-funded by the Research Directorate-General of the European Commission within the Research and Technological Development (RTD) activities of the FP7 Thematic Priority Environment.

1. Introduction

using wavelet transforms were assessed (chapter 4). Due to the project guidelines, we focused on two soil parameters: organic carbon (C_{org}) and total nitrogen (N_{tot}). A mobile platform to conduct on-the-go measurements directly in the field using a portable spectrometer was constructed and tested to estimate C_{org} and N_{tot} (chapter 6). To better understand the influence of soil moisture on the spectra collected in the field, the reflectance of moist soil samples was investigated in the laboratory (chapter 5). The main results and conclusions are summarised in chapter 7. The used methodologies and their theoretical backgrounds are lined out in chapter 2, which possibly can serve as 'rough guide' to VNIRS soil analysis.

PS: Unfortunately, Twain's advice is followed by investors like speculative trusts, enterprises, and rich countries nowadays more than ever. During the last years, vast areas of arable land – especially in Africa – were bought or leased for long-terms to establish huge plantations cultivating profit-yielding products like export goods and biofuel. Amongst other, expulsion of local residents, social conflicts, and severe ecological problems are consequences of the so-called land grabbing. According to estimations of the International Food Policy Research Institute (Ifpri), 20 millions of hectares of African soil was sold or long-term leased between 2008 and 2010 (Baxter, 2010).[3]

---

[3] Available online: http://www.monde-diplomatique.fr/2010/01/BAXTER/18713 (French; for translations use search functions of http://mondediplo.com and http://www.monde-diplomatique.de)

2. Methods

2.1. Principles of Visible and Near Infrared Spectroscopy (VNIRS)

Electromagnetic waves irradiated to a material are transmitted, reflected, and/or absorbed. For opaque materials like soil, no transmittance occurs. Thus, a portion of the energy is absorbed and the remaining is reflected. The proportions of absorption and reflection are wavelength dependent and provide information about the physical and chemical properties of the material. Two types of reflectance are observed: specular reflectance and diffuse reflectance. The latter is measured by VNIRS. For specular reflectance, the angle of reflection is equal to the angle of incidence (figure 2.1). At all other angles, only diffuse reflectance is observed. For materials with matte finish, e.g. white paper, powders or soil, specular reflectance is very low (Siesler, 2008). To describe the radiation flux I_r remitted by diffuse reflection, the Lambert cosine law was proposed:

$$\frac{dI_r/df}{dw} = \frac{CS_0}{\pi}\cos\alpha\cos\vartheta = B\cos\vartheta \tag{2.1}$$

where df represents the size of the reflecting area (cm^2), dw represents the solid angle in steradians (sr), C is a (material-dependent) constant representing the fraction of incident radiation flux remitted, S_0 is the irradiation intensity (W cm^{-2}) for normal incidence, and α and ϑ are the angles of incidence and view, respectively. Most variables can be summarised by B (W cm^{-2} sr^{-1}) representing the surface brightness or radiation density (Siesler, 2008). Because all variables except C are kept constant for spectroscopic measurements, the differences in the spectra express differences related to the composition and the surface properties of the measured materials and can therefore be used for chemical analyses.

The total absorption of a specific material comprises a physical and a chemical component. The physical component is mainly caused by the geometry of the measured surface (particle size,

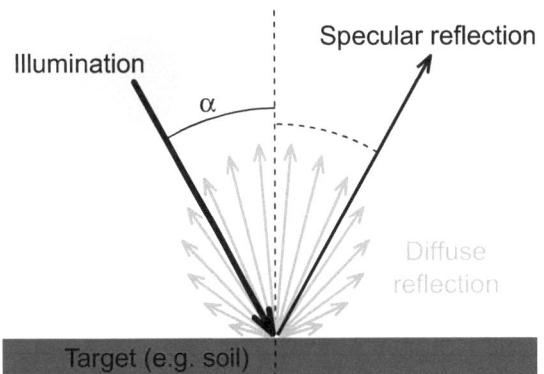

Figure 2.1 Schematic of specular and diffuse reflection. The angle of specular reflection corresponds to the incidence angle α, whereas diffuse reflection is observed for any angle of view $\vartheta < 90°$.

2. Methods

Table 2.1 Characteristics of the visible, the near infrared (NIR), and the mid infrared (MIR) ranges of electromagnetic radiation.

	Visible	NIR	MIR
Wavelength (nm)	400 - 760	760 - 2500*	2500* - 25 000
Wavenumber (cm^{-1})	25 000 - 13 200	13 200 - 4000	4000 - 400
Absorptions by			
Fundamental vibrations			X
Overtones and combinations		X	X
Electronic processes	X	X	
Related to	Electrons	CH/OH/NH functionalities	Polar functionalities
Absorption amplitude		weak	strong
Absorption characteristics		broad & overlapping	sharp bands
Used soil samples		sieved < 2 mm (ev. milled)	finely ground (or diluted in pressed KBr)

*Some references defined 3000 nm as limit between NIR and MIR.

shape of particles, ...), whereas the chemical component is related to the composition of the material. The nature of the occurring absorptions depends on the measured wavelength range. The Mid Infrared (MIR) including radiation of wavelengths from 2500 to 25 000 nm and the Visible and Near Infrared (VNIR) including wavelengths from 400 to 2500 nm both can be used for soil analyses.[1] Different properties of the measured materials are targeted by the two ranges, and it depends on the intention of the analyses which of the two is superior.

2.1.1. Physical absorption mechanisms

The material's particle size strongly influences its reflection. Bowers & Hanks (1965) reported a 'rapid exponential increase in reflectance with decreasing particle size' for samples of clay minerals differing in particle size only. They attributed this effect to the increased mass density for smaller particles: smaller particles are more densely packed what implies an elevated concentration of the measured material. In addition, the porosity and thus the pathlength (the distance that the light beams are covering) are decreased.

Surface roughness on a micro-scale was supposed to be the determining factor in explaining reflectance changes in respect of particle size: finer particles generated smoother surfaces, whereas coarse, irregularly shaped aggregates produced a complex surface comprising a large number of inter-aggregate spaces (Orlov, 1966, in Baumgardner et al., 1985; Bowers & Hanks, 1965). These may act as light traps and extinguish a good part of the irradiation. Surface roughness and mass

---

[1] Due to historical reasons, MIR spectroscopists are using the unit *wavenumber* $\tilde{\nu}$ *(cm^{-1})* in their spectra, whereas VNIR spectroscopists usually are using *wavelength* λ *(nm)* instead.

$$\tilde{\nu} = \lambda^{-1} \qquad \text{Conversion factor nm} \leftrightarrow \text{cm}^{-1}: 10^7$$

2.1. Principles of Visible and Near Infrared Spectroscopy (VNIRS)

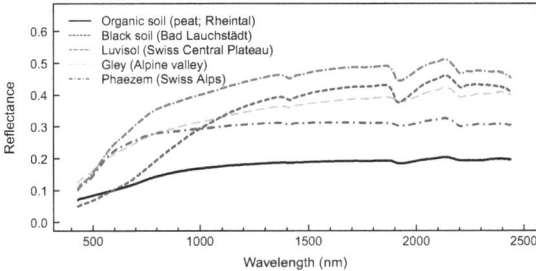

Figure 2.2 Reflectance spectra of five randomly selected soil samples in oven dry condition (four samples from the *nabo* set and one from the *lauch* set; absorbance spectra of same samples are provided by figure A.10, page 113).

density also differ considerably between distinct soils and induce spectral differences between them.

Most of the reflectance increase was observed for particles smaller 0.4 mm. In contrast, only slight differences were observed between fractions of bigger particles (Bowers & Hanks, 1965). It was assumed that only the clay fraction can be assessed by VNIRS, whereas the other soil texture fractions have little influence on the spectra.[2]

The penetration depth of VNIR radiation is an important aspect, too. It governs the volume sampled during spectroscopic measurements. Usually, the penetration depths for VNIR are assumed to range from micrometers to few millimeters. For a microcrystalline cellulose powder (particle size ranging from 65 to 300 μm), penetration depths ranging from 0.3 to 2.5 mm were reported. The penetration depth was strongly wavelength dependent, it was clearly higher for wavelengths below 1500 nm and highest between 600 and 1000 nm (Siesler, 2008).

Most often, soil samples measured in the laboratory by VNIRS are dried, passed through a 2 mm-mesh, and measured without further treatments, although sometimes they are also milled. Contradictory results were reported whether milling the samples reduces estimation errors, but the differences seem to be small, generally (Stenberg et al., 2010). The effect of milling depends on the targeted soil property, as it might reduce on one hand errors related to surface roughness, on the other hand soil roughness might be an indicator for the assessed soil property. In addition, breaking up the soil aggregates makes the inside of them visible to the spectrometer. In contrast to VNIRS, finely ground samples are a prerequisite for MIR spectroscopy because longer wavelengths are more sensitive to soil roughness. It is a big advantage of VNIRS over MIRS that milling is not necessary.

2.1.2. Chemical absorption mechanisms

In the VNIR, there are two relevant absorption processes of chemical nature: electronic processes and vibrational processes (Hunt, 1977).

Electronic processes Several electronic processes – including crystal-field effects, charge transfer, colour centres, and conduction band transitions – induce absorptions in the visible and the NIR. Simplifying, atoms in some molecules may exist in different energy states. When they

---

[2] This effect is boosted by the fact that clay minerals (which dominate the clay fraction) additionally exhibit chemical absorption features in the VNIR, but quartz (the main constituent of the sand fraction) shows no absorption in that range.

2. Methods

Table 2.2 Important VNIR absorption bands observed for soil samples (Hunt, 1977)

Wavelength (nm)	Explanation
Water H_2O	
1400-1500	Combination of symmetric and asymmetric OH-stretch
1900-2000	Combination of H-O-H bend with asymmetric O-H stretch
⇒ *If water molecules are present, both bands are observed* ***together***.	
2600-2800	The absorption maxima of the fundamentals of symmetric and asymmetric OH-stretch fall into the MIR, but their influence is also visible in the NIR as decline in reflectance beyond 2200 nm.
Hydroxyl OH	
1400-1500	Combination of symmetric and asymmetric OH-stretch
⇒ *Same as for H_2O, but no absorption band around 1940 nm.*	
2200-2400	Hydroxyl groups attached to metal, paired absorption peak, exact position depends on metal and structure of mineral.
Carbonates	
1900, 2000, 2160, 2350, 2550	Combinations and overtones of C-O bonds; the two last are reported to be clearly doubled and more intense than the remaining.
Quartz	
(none)	Si-O bonds possess strong absorptions in the MIR, but not in the VNIR.

change between them (referred to as transition), electromagnetic radiation is absorbed or emitted at specific wavelengths. Electronic features in the VNIR are most often caused by the presence of iron (Fe^{3+}, Fe^{2+}), but also features of other transition elements like nickel, copper, manganese, and chromium occur. Other prominent examples are fluorite minerals absorbing in the visible and provoking very intense colours (Hunt, 1977, for a comprehensive overview on electronic processes see therein).

Vibrational processes Molecules consist of single atoms held together by chemical bonds. Atoms are periodically moving within the molecules, e.g. they are stretching (the length of the bond is changed) and bending (the angle between two bonds is changed). See Workman & Weyer (2007) for a comprehensive overview on molecular vibrations. Atoms are excited to vibrate by electromagnetic radiation. This process absorbs energy, for this reason less energy is remitted than incident. Energy is absorbed at the wavelengths corresponding to the frequencies of the molecular vibrations, and that is the reason why absorption bands are visible in spectra. Fundamental vibrations exhibit frequencies roughly from 10^{12} to 10^{14} Hz corresponding to 30 000 to 3000 nm wavelengths.[3] Depending on the molecular structure, absorption due to overtones (multiples of fundamental vibrations) and combinations of them occur at higher frequencies, but they are much weaker (by a factor of 10 to 100 times going from a vibration to the next overtone). Only overtones and combinations of groups with very high fundamental frequencies are visible in the NIR, the most important of them are O-H, C-H, and N-H groups (Hunt, 1977).

---

[3] Wavelength $\lambda = c/f$ with c representing speed of light ($3.0 \cdot 10^8$ m s^{-1}) and f representing frequency.

2.1.3. Amplitude of absorption

Because the intensity I of diffuse reflection depends on the angle and the distance of the observation relative to the reflecting surface (cf. equation 2.1), VNIRS measurements are compared to a so-called *white reference* measurement which was taken from the same angle and distance, but using a surface exhibiting 100 % diffuse reflectance. The intensity of the white reference measurement is assumed to be identical to the incident radiation (the radiation before the interaction with the surface), and is denoted I_0. Reflectance (R) is defined as the ratio of I and I_0 (Workman & Weyer, 2007).

The frequencies of molecular vibrations determine at which wavelengths absorptions occur, while the amplitude of the absorption is assumed to be proportional to the concentration c of that molecule, the pathlength l which describes the thickness of the sample interacting with the light beams, and the absorptivity ε which depends on the molecule and the matrix containing it. The amplitude of absorption is described by the Bougner, Lambert, and Beer relationship which is very often called Beer's law (Workman & Weyer, 2007):

$$R = \frac{I}{I_0} = 10^{-\varepsilon c l} \tag{2.2}$$

Based on this assumption, reflectance data are usually log-transformed to absorbance (A):

$$A = -\log\left(\frac{I}{I_0}\right) = -\log R = \varepsilon c l \tag{2.3}$$

Some scientists and books (and also this work) are using the natural logarithm for equation 2.3, while others prefer the decadic logarithm. Both approaches are feasible, but they should not be messed, because different values for ε result.

If absorption bands can be clearly attributed to a certain constituent of the measured material, equation 2.3 can be used to establish linear calibrations based on one or few selected wavelengths. Unfortunately, absorption bands of soil in the VNIR are broad, overlapping, and mostly weak. In addition, soil is a very complex and variable matrix. Therefore, statistical methods are needed for most soil properties to detect the relevant wavelength ranges and to establish calibrations to estimate them (cf. section 2.5). However, a calibration is always necessary for VNIRS analyses, it cannot be used as stand-alone technique.

2.2. Soil samples and archives

Various sets of soil samples were used for calibration and validation. If not stated otherwise, dried and sieved (< 2 mm) samples were used. In addition to samples collected and analysed by the University of Bern, archived soil samples stored by different institutions were used. These samples had already been analysed by a reference method. The results of these analyses were provided by the owners of the archives and could be used for data analysis. Within the present work, archived samples of various Swiss institutions were measured by VNIRS, including the Swiss Soil Monitoring Network (NABO), Zürich-Reckenholz, the University of Applied Sciences Changins (École d'Ingénieurs de Changins, EIC), Nyon, the cantonal soil monitoring units of Fribourg (FRIBO; these samples had been measured by Hauert, 2007, for a previous project) and Solothurn (Fachstelle Bodenschutz), and the Swiss Federal Institute for Forest, Snow and

2. Methods

Figure 2.3 Used laboratory measuring setup: FieldSpec 3 spectrometer, muglight, and petri dish containing soil material placed atop of the muglight for the spectral measurement.

Landscape Research (WSL), Birmensdorf. Thanks to this collaborations, a lot of time and effort could be saved. The used sets of soil samples for each chapter are described therein.

2.3. VNIRS laboratory measurements

2.3.1. Equipment and material

Two portable spectroradiometers manufactured by Analytical Spectral Devices Inc. (ASD, Boulder, CO, USA) were used; a *FieldSpec Pro* and a *FieldSpec 3 High Resolution*. Both devices are equipped with three detectors: one 512 element silicium photodiode array covering 350 to 1000 nm wavelengths, and two indium gallium arsenide photodiodes covering 1000 to 2500 nm wavelengths. The change between the two latter sensors occurs at 1800 nm for the FieldSpec Pro and at 1830 nm for the FieldSpec 3. The sampling interval of the spectrometers is 1.4 nm for wavelengths below 1000 nm and 2 nm above 1000 nm. The data are interpolated by the spectrometers to 1 nm intervals. A spectral resolution of 3 nm at 700 nm, 8.5 nm at 1400 nm, and 6.5 nm at 2100 nm was reported for the FieldSpec 3 (ASD, 2009). For the FieldSpec Pro, lower resolutions between 10 and 12 nm were reported for wavelengths above 1000 nm. Seiler (2006) determined the signal-to-noise ratio for the used FieldSpec Pro device and recommended to discard wavelengths below 430 nm and above 2440 nm. This recommendation was also used with the FieldSpec 3 device in order to receive spectra covering the same wavelength range for both spectrometers.

A muglight contact probe was used to measure the reflectance of soil samples through the bottom side of petri dishes. The muglight was equipped with an 12 V tungsten quartz halogen lamp (4 W, 2900 K colour temperature) for illumination. The measured samples were illuminated through a sapphire window. The spectrometer's optical interface, which has a field of view of about 25°, was placed adjacent to the halogen lamp. The collected radiation was sent through a fiber optic cable to the spectrometer. The spot size captured by the measurements was a nearly circular area with a diameter of 12 mm (ASD, 2011).

Schott Duran petri dishes (Duran Group, Mainz, Germany) were used for the spectroscopic measurements. They are made of borosilicate glass exhibiting very low absorption in the VNIR (Seiler, 2006). Round petri dishes of 6 cm diameter were used for the soil samples, and a 10 cm diameter petri dish was used for the Spectralon panel. The softwares provided by ASD (RS^3 and ViewSpec Pro) were used for data collection and export.

2.3.2. White reference

White reference measurements are needed to convert the measured radiance to relative reflectance. For this purpose, a white reference standard is measured by the spectrometer using exactly the same setup (geometry, illumination, ...) as for the common measurements. The preferred standard for VNIR spectroscopy is Spectralon which consists of compressed polytetrafluoroethylene powder and is an almost perfect Lambertian reflector (Voss & Zhang, 2006). This material exhibits a very high diffuse reflectance within the entire VNIR: a reflectance of approximately 0.99 is observed from 400 to 1900 nm wavelengths radiation, and from 1900 to 2500 nm diffuse reflectance still exceeds 0.96 (Weidner & Hsia, 1981; Goldstein et al., 2003).

The used white reference panels should be handled with care because using a degraded panel results in corrupt spectral measurements. Spectralon panels can be contaminated by dust (e.g. from filling-in the soil samples) and sweat resulting in decreased reflectance. Furthermore, ageing of Spectralon due to radiation also decreases its reflectance. Ageing effects were even observed for low radiation intensities, but seemed not to be dramatic for wavelengths above 350 nm, whereas ultra-violet radiation induced reflectance decreases as high as 1 % within several hours of irradation (Möller et al., 2003). Therefore, Spectralon panels should be regularly tested by another reference or internal standard samples.

Within the present work, a round Spectralon of 10 cm diameter was used. It was placed in a petri dish for the white reference measurements. A second, identical Spectralon was used as control. It was stored sealed in a save place and exclusively used to monitor the reflectance of the Spectralon used for the usual laboratory work. From time to time, the reflectance of the two panels were compared by using the control panel as white reference and measuring the used panel in reflectance mode. The resulting reflectance curve was used to correct the measured spectra for the contamination of the used panel:

$$R(\lambda)_{\text{corrected}} = \frac{R(\lambda)_{\text{measured}}}{R(\lambda)_{\text{used Spectralon}}} \qquad (2.4)$$

2.3.3. Measuring procedure

The procedure presented in this section and in figure A.2 (page 104) was used for oven and air dried samples that were sieved using a 2 mm mesh.

Spectrometer and muglight were switched-on and warmed-up for on hour if possible, but at least for 30 minutes. Thereafter, a petri dish containing the Spectralon was placed at the muglight to optimise the optical settings of the spectrometer by the built-in routine of the RS^3 software. The software was set to collect 20 co-added scans for every measurement and white reference, and to use 30 co-added scans for the dark current measurements. Prior to measuring the first soil sample, the petri dish containing the Spectralon was placed at the muglight to take a white reference measurement. This step was repeated every five to ten minutes and after breaks. Then, soil samples were mixed well and 15 to 20 g of soil material was filled into a petri dish. The petri dishes were gently moved to ensure that the material was settled compactly. The petri dishes were placed at the muglight and two spectra were collected for every sample; between the two measurements, the petri dish was rotated by 90°. After measuring all samples, the spectral data were exported as reflectance data to an ASCII-file.

2.4. Reference analyses

VNIR spectroscopy cannot be used as stand-alone analytical method. Measurements by established reference methods are required to derive calibrations and to validate them. The reference methods used for the present work are described within this section.

2.4.1. Total and organic carbon by dry combustion

Soil carbon is present in organic and inorganic forms: organic carbon (C_{org}) is the main constituent of organic matter (OM) which includes the living biomass, dead biomass in different states of degradation, and its stable decay products in the form of humus. Inorganic forms are mostly present as carbonate minerals (Nelson & Sommers, 1996). Elemental forms of C, e.g. charcoal, are not relevant for most soils (Schumacher, 2002). The total carbon (C_{tot}) is efficiently determined by dry combustion methods. Their principles are simple: a small portion of the sample is burned in a closed chamber in a stream of purified O_2. All carbon in the sample is oxidised to CO_2 which can be determined in the effluent gas stream by different methods, e.g. by conductimetric, spectrophotometric, and volumetric procedures (Nelson & Sommers, 1996). For soil samples containing only C_{org}, the results of dry combustion can be considered as equal to C_{org} contents. Otherwise, inorganic carbon must be removed prior to dry combustion (e.g. by adding sulfuric acid) or must be determined by another method and subtracted from the C_{tot} estimates in order to receive C_{org} contents.

2.4.2. Organic carbon/matter by the Swiss standard method

Usually, OM contents are not determined directly; instead the C_{org} content is measured and multiplied by 1.725 (FAL, 1996). Dichromate digestion techniques are widely used to estimate soil C_{org} and OM, and numerous versions of them exist, the most prominent being the one presented by Walkley & Black (1934). Also the Swiss standard method according to FAL (1996) falls into this category. It was used by some of our partners for OM analyses. Potassium dichromate ($K_2Cr_2O_2$) and concentrated sulphuric acid (H_2SO_4) are added to the soil sample in order to oxidise the carbon therein to carbon dioxide (CO_2). An excess of potassium dichromate is added and the amount not used for oxidation is determined by back-titration. Based on the amount of potassium dichromate used for oxidation, the amount of C_{org} in the sample can be estimated.

Depending on the stability of the present carbon forms and the used method, varying fractions of organic carbon are oxidised. Therefore, discrepancies were found between C_{org} determined by dry combustion and dichromate oxidation methods. According to Nelson & Sommers (1996), organic carbon recovery rates from 27 to 144 % compared to dry combustion were reported for the Walkley-Black method. Comparing organic carbon contents determined by the Swiss standard method and dry combustion of eight samples originating from the Swiss Central Plateau (Reisiswil and Gondiswil, Canton of Bern, samples provided by Ruth, 2010) indicated a recovery rate of 79 % (figure 2.4). This value was used throughout this work to compare results of the two analytical methods knowing that this value was not necessarily correct for all Swiss soils.

2.4.3. Total nitrogen

Analogically to carbon, nitrogen is present in organic forms (as constituent of organic matter) and inorganic forms (ammonium, nitrate, nitrite). Total soil nitrogen is usually determined by dry

2.5. Calibration

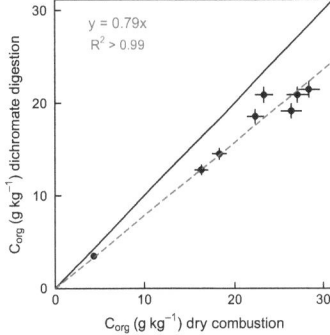

Figure 2.4 Comparison of C_{org} contents determined by dry combustion (CN analyser) and dichromate digestion according to the Swiss standard method.

combustion. The principles are very similar to those of total carbon: a small portion of the sample is burned in a closed chamber in a stream of purified O_2. Nitrogen is oxidised to nitrogen oxides which subsequently are reduced to N_2, e.g. by passing them through a column consisting of copper. The N_2 in the effluent gas stream then can be determined by different methods (Bremner, 1996).

2.4.4. Equipment and procedure for CN analyses by the University of Bern

All analyses for total C and N by the University of Bern were conducted using a Vario EL III Element Analyzer (Elementar Analysensysteme GmbH, Hanau, Germany) in CN modus. A subsample was taken from the mixed soil samples and finely ground in a mortar. Between 100 and 200 mg of soil material was packed into small tin containers. These were burned at temperatures near 1000°C by the CN analyser in a stream of helium and oxygen. Thus, all forms of carbon and nitrogen were oxidised to CO_2 and NO_2 and NO, respectively. The nitrogen oxides were subsequently reduced to N_2 in a column containing copper, and then CO_2 and N_2 were separated by a column absorbing CO_2 in cold state and desorbing CO_2 when heated. The quantities of CO_2 and N_2 in the effluent gas stream were determined by a detector measuring the thermal conductivity of the gas stream and put in relation to the weight of soil material fillediin. Samples of pure glutamic acid were used as standard to correct for systematic errors in the measurements. See section 3.4 for considerations about accuracy and reproducibility of the used technique.

2.5. Calibration

Because the absorption features in the VNIR are weak, broad, and overlapping, statistical methods are used the detect wavelength ranges relevant to estimate a specific (soil) parameter. Furthermore, various techniques are used to pre-treat the spectral data prior to data analysis. The field including these activities is designated as *chemometrics*. The topics of this area relevant for the present work will be presented within this section.

2.5.1. Pre-treatment of spectral data

The used VNIR spectrometers compared the measured radiance per wavelength and soil sample to a precedent white reference measurement, and reported the results as reflectance per wavelength:

2. Methods

$R(\lambda)$. Reflectance values usually vary from zero to one and represent the amount of energy reflected at every wavelength. To detect and remove bad measurements, the spectra should first be screened visually. Remember that two spectra were recorded for every soil sample, so if only one spectrum exhibited errors, the remaining was maintained. If the spectra of a sample were very different, but none was obviously incorrect, the measurements were repeated or deleted.

Next, corresponding spectra were merged into one spectrum by calculating their mean for every wavelength. Then, the steps occurring at the sensor changes of the spectrometer (1000 and 1800 nm for FS Pro; 1000 and 1830 nm for FS3) were removed by shifting the right hand side of the spectra to match the left hand side. For this purpose, the medians of the first four wavelengths at every side of the sensor changes were compared. Then, a Savitzky-Golay smoothing filter (Savitzky & Golay, 1964) of length five was applied before removing wavelength ranges with low signal-to-noise ratios (below 430 nm and above 2440 nm).

Only every fourth data point was kept (430, 434, ..., 2438 nm) in order to reduce the size of the datasets. The wavelength interval of four nm was chosen to be slightly smaller than the optical resolution achieved by the spectrometer. The pre-treated reflectance data was used for data analysis, or was further transformed to absorbance (A)

$$A(\lambda) = -\log R(\lambda) \tag{2.5}$$

which seems more adequate for data analysis assuming that Beer's law applies. Derivatives of reflectance data were calculated applying appropriate Savitzky-Golay filters of length 25 instead of the smoothing filter. The process of pre-treating spectral data is also described in figure A.3 (page 105).

2.5.2. Pre-screening of spectral data

Before calibrations can be established, the used datasets must be screened to detect spectral outliers and potential sub-groups within the dataset. The most straightforward way to do this is principle component analysis (PCA) of reflectance or absorbance spectra. Principle components are linear combinations of the dataset's variables that are determined by their ability to account for the variability in the dataset (Naes, 2002). Therefore, the first principle components represent most of the variability, and usually few components account for more than 90 % of the total variability. Projecting the observations of the dataset onto the derived components yields the 'new' values related to this components and are called *scores*.

Outliers can be detected by screening scatter plots of the scores of the first principle components or by calculating the distance of every observation from the mean of all observations. The distances are calculated as *Mahalanobis distance (MD)* using the scores of the first to the a-th principle component. They represent distances within the score space and are therefore called *score distances* (*SD*; Hubert et al., 2008). For a p-dimensional dataset X consisting of n observations

$$X = \begin{pmatrix} x_{1,1} & x_{1,2} & \cdots & x_{1,p} \\ x_{2,1} & x_{2,2} & \cdots & x_{2,p} \\ \vdots & \vdots & \ddots & \vdots \\ x_{n,1} & x_{n,2} & \cdots & x_{n,p} \end{pmatrix} \tag{2.6}$$

2.5. Calibration

the *MD* of observation $x_i = (x_{i,1}, x_{i,2}, \ldots, x_{i,p})$ from the centre of X is defined by

$$MD_i = \sqrt{(x_i - \bar{x})^T \Sigma_X^{-1} (x_i - \bar{x})} \tag{2.7}$$

with $\bar{x}$ being the (column-wise) mean vector of X, and Σ^{-1} being the inverse of the covariance matrix of X which is defined by (De Maesschalck et al., 2000)

$$\Sigma_X = \begin{pmatrix} \text{cov}(X_1, X_1) & \text{cov}(X_1, X_2) & \ldots & \text{cov}(X_1, X_p) \\ \text{cov}(X_2, X_1) & \text{cov}(X_2, X_2) & \ldots & \text{cov}(X_2, X_p) \\ \vdots & \vdots & \ddots & \vdots \\ \text{cov}(X_p, X_1) & \text{cov}(X_p, X_2) & \ldots & \text{cov}(X_p, X_p) \end{pmatrix} \tag{2.8}$$

where $X_1, \ldots, X_p$ represent the columns of X. The elements on the diagonal of the matrix represent the variances of the columns of X, because $\text{cov}(X_i, X_i) = \text{var}(X_i)$.

It is obvious that *SD* depends on a, the number of included principle components. For observation t_i of a given score matrix T_a containing a principle components, the *SD* is calculated by

$$SD_{i,a} = \sqrt{(t_{i,a} - \bar{t}_a)^T \Sigma_{T_a}^{-1} (t_{i,a} - \bar{t}_a)} \tag{2.9}$$

with $\bar{t}_a$ being the (column-wise) mean vector of T_a, and Σ^{-1} being the inverse of the covariance matrix of T_a.

Because the squared *SD* are approximately χ^2-distributed, the quantile $\chi^2_{a, 1-\alpha}$ of the chi-squared distribution with a degrees of freedom (where α is the level of significance) can be used as threshold value (Hubert et al., 2005). Observations with $SD^2 > \chi^2_{a, 1-\alpha}$ are considered outliers and removed from the dataset. PCA is very sensitive to outliers and can be heavily influenced by them (Hubert et al., 2008). Therefore, it is recommended to start with few principle components (two or three) and high significance levels (0.02 or 0.01) to remove the most extreme outliers only. This procedure is then iterated by recalculating PCA for the reduced dataset. During the iterations, the number of used principle components and possibly the significance level can be increased until no clear outliers remain in the dataset. The number of components finally included and the significance level depend on the used dataset and should be specified by the user consulting plots of *SD* and principle components. Based on my experience, I recommend using five to ten components and a significance level of 0.02 or 0.05. Remember that the described process is intended to remove clear spectral outliers prior to regression analysis, but as the regression analysis itself again includes outlier removal, we do not need to care to much whether a specific sample should be removed or not. The whole process is also described by the flow chart in figure A.4 (page 106).

2.5.3. Linear regression models

Linear regression models can be used to estimate a vector y of n observations by p predictor variables represented by column vectors $x_1, x_2, \ldots, x_p$ (Hastie et al., 2009):

$$y = X\beta - d = \beta_0 + \beta_1 x_1 + \beta_2 x_2 + \ldots + \beta_p x_p - d \tag{2.10}$$

with $X = (1, x_1, x_2, \ldots, x_p)$ where 1 represents a vector of length n containing ones only, $\beta = (\beta_0, \beta_1, \ldots, \beta_p)^T$ representing the (unknown) parameters or coefficients of the model, and vector d

2. Methods

being the residuals of the model that are assumed to follow a normal distribution. The parameters β can be estimated by the method of *least squares* minimising the sum of squares of the residuals $d = \hat{y} - y$ using the following equation:

$$\beta = (X^T X)^{-1} X^T y \qquad (2.11)$$

The linear regression model estimates then the observations of y by:[4]

$$\hat{y} = X\beta = \beta_0 + \beta_1 x_1 + \beta_2 x_2 + \ldots + \beta_p x_p \qquad (2.12)$$

Any unknown observation(s) X_u may be estimated by $\hat{y}_u = X_u \beta$.

Linear regression models are inappropriate to be applied directly to spectral data, because these are consisting of a large number of strongly correlated predictor variables. The resulting models may be very unstable, and its estimates of poor performance (Naes, 2002). Therefore, the number of predictor variables must be reduced, either by selecting a small subset of variables (wavelengths) for the regression, or by compressing the data and using the compressed data in the linear regression. For spectral data, the first strategy only seems adequate, if specific wavelengths influenced strongly by the soil property of interest can be identified, e.g. for estimating soil water contents (cf. chapter 5). For most soil properties, VNIR absorption bands are weak, broad, and overlapping (Workman, 2008). In these cases, compressing the spectral data – e.g., by partial least squares regression or wavelet decomposition; see following sections – is more efficient. The relevant ranges of wavelengths are then selected statistically, and not manually based on spectroscopic theories.

2.5.4. Partial least squares regression

Partial least squares regression (PLSR) resolves the collinearity problem described above by deriving 'new' variables as linear combinations from the original data and using these new variables in a linear regression model. The derived variables are orthogonal and thus not correlated, and a small number of them is sufficient to estimate the original data X satisfactorily. The dimensionality of the dataset is strongly reduced, because only few linear combinations representing most of the relevant information are maintained. PLSR can be interpreted as projection to latent structures (Naes, 2002; Wold et al., 2001; Bjørsvik & Martens, 2008).

The derived variables are called *PLS factors*, whereas the projection (the values) of the original data points on the PLS factors are called *scores*. Basically, PLSR can be described by the following equations:

$$T = XW \qquad (2.13)$$
$$X = TP^T + E \qquad (2.14)$$
$$y = Tq + f \qquad (2.15)$$

Matrix X consists of p column vectors $(x_1, x_2, \ldots, x_p)$ representing predictor variables of n observations each, and vector y represents the variable that should be estimated.[5] For PLSR, X and y

---
[4] Throughout this work, y represents reference values ('true' values), and $\hat{y}$ represents estimates thereof by a regression model.
[5] Here, formulas for a univariate variable y are presented. PLSR also handles multivariate regression models with

2.5. Calibration

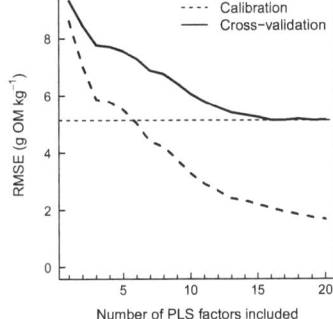

Figure 2.5 RMSE as a function of the number of PLS factors used for the 'fr-so' model to estimate organic matter presented in section 3.1

are mean-centred (from every variable, its mean is subtracted). The scores $T = (t_1, t_2, \ldots, t_a)$ are calculated by applying the weights W (in some references denoted as R) on X. For the inverse step, matrix P containing the so-called *X-loadings* is required. The *y-loadings* q link T and y. E and f represent the residuals that are attributed to noise and unimportant information. Once a PLSR model is established, regression coefficients β can be derived, and the model can be described by:

$$y = XWq + f = X\beta + f \tag{2.16}$$

But how are the PLS factors obtained? An iterative procedure is applied: For the first PLS factor, the covariance between y and all possible linear functions of X is maximised. Then, the scores for the first PLS factor can be calculated. The first factor is subtracted from X and y which are subsequently used to derive the second PLS factor applying the covariance criterion again. The procedure is repeated until $a \leq p$ PLS factors are derived. In other words, each factor is calculated from the residuals (E and f) of the previous factors (for details see Naes, 2002).

The optimum number a of PLS factors should be determined assessing the cross-validation RMSE (equation 2.22). While calibration RMSE continuously decreases when including more PLS factors, cross-validation RMSE starts to increase again above a certain number (figure 2.5). Usually, as few PLS factors as possible should be included ($a \ll p$) to avoid over-fitting.

PLS factors are calculated similarly to principle components, except that for the first, covariance in (X, y) is maximised, whereas for the latter, variance in X is maximised. Therefore, PLS factors are more directly related to the variability in y than principle components. Usually, the performance of PLSR and regression on principle components is comparable, although the latter requires more components to achieve its maximum performance (Naes, 2002).

All calibrations for the present work were calculated using the SIMPLS algorithm (de Jong, 1993). It was proposed as fast PLSR algorithm. For univariate y, the results are identical to those of classical PLSR, whereas for multivariate Y, slight differences were observed. The complete algorithm is provided in section A.1.1 (page 111).

$Y = (y_1, y_2, \ldots, y_m)$; see cited references for details.

2.5.5. Diagnostic plots for PLSR

Diagnostic plots are helpful tools to interpret the derived models and to remove outliers. Partial least squares regression is very sensitive to outliers, and few outliers within the dataset can completely alter the resulting model and corrupt its performance (Hubert et al., 2008). Both steps of PLSR – computing the PLS factors and the subsequent fitting of a linear regression model – exhibit very low resistivity against outliers. Therefore, outliers should be detected and removed from the dataset and the model recalculated. As some outliers might have remained undetected because more extreme outliers were present, the process of outlier removal described below should be iterated (cf. figure A.4).

As described above, PLSR shrinks a high-dimensional dataset into a dataset of few dimensions. The observations of the original dataset are projected onto a subspace within the original space. During this step, some information in our dataset gets lost. In respect of the whole dataset, the information loss is small, but for some samples it might be big, especially for samples very different from the majority. The *orthogonal distance (OD)* as defined by Hubert et al. (2008) represents the Euclidean distance between an observation and its projection onto the subspace and can be used as proxy for the information loss. It corresponds to the root of the squared residuals of the observation's reconstruction by the used PLS factors (cf. equation 2.14). For observation $x_i = (x_{i,1}, x_{i,2}, \ldots, x_{i,p})$ from a p-dimensional dataset X and a given model including a PLS factors, it is calculated by

$$OD_{i,a} = \|x_i - \bar{x} - t_{i,a} P_a\| = \sqrt{(x_i - \bar{x} - t_{i,a} P_a)^T (x_i - \bar{x} - t_{i,a} P_a)} \quad (2.17)$$

with $\bar{x}$ representing the (column-wise) mean vector of X (which is required because PLSR implies mean-centring), $t_{i,a}$ representing the scores of observation x_i, and P_a the loadings of the model. The product of $t_{i,a}$ and P_a corresponds to the observation's reconstruction.

The locations of the observations *within* the subspace produced by PLSR are reflected by the scores T. To assess the distances between the observations within this subspace, score distances (SD) are calculated similarly as for PCA scores (cf. section 2.5.2):

$$SD_{i,a} = \sqrt{(t_{i,a} - \bar{t}_a)^T \Sigma_{T_a}^{-1} (t_{i,a} - \bar{t}_a)} \quad (2.9)$$

Both SD and OD are calculated for the cross-validation estimates. On one hand, outliers are more promoted in cross-validation than in calibration – and on the other hand, the distances of the cross-validation estimates are more suitable for comparison with those of unknown samples (see below).

After defining OD and SD, we are ready for the first diagnostic plot: a scatter plot of the two serves as outlier map (figure 2.6, left side). While most observations are located in the bottom left quarter of the plot, few samples appear outside of it. These samples are outliers in respect of SD or OD, or both. The separation between outliers and 'good' observations can be done using thresholds or simply by eye. The latter approach is easier and usually efficient, because it is not clear what threshold to use for OD whose underlying distribution remains unclear (Hubert et al., 2008). For SD, $\chi^2_{a,1-\alpha}$ should be used (cf. section 2.5.2).

The leverage of an observation reflects its influence on the regressed model. Influential observations possess high leverages. For the SIMPLS algorithm, the leverages h for n observations are

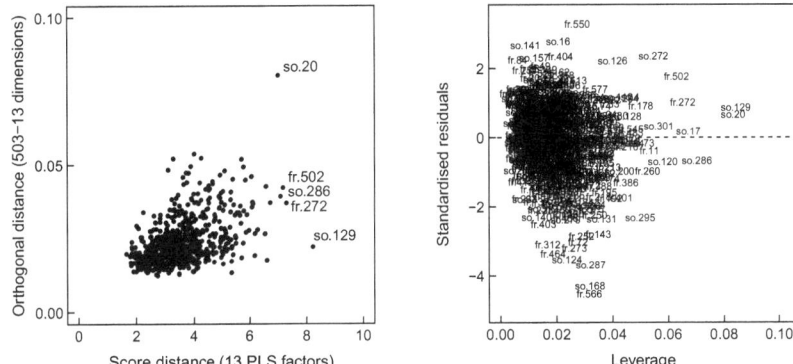

Figure 2.6 SD-OD-plot (left) and leverage plot for the 'fr-so' model to estimate organic matter presented in section 3.1

estimated from the score matrix T by

$$h = \mathrm{diag}(T_a T_a^T) + 1/n \qquad (2.18)$$

where 'diag' extracts the diagonal of the matrix (de Jong, 1993). Plotting h against the standardised residuals brings out the leverage points within the regression, and – as visible in figure 2.6 – these usually correspond to the outlying observations in the SD-OD plot.

Once the outlier removal has been finished, other common diagnostic plots like the Tukey-Anscombe plot (residuals vs. fitted values), the scale-location plot (squares of absolute residuals vs. fitted values), and the normal-QQ plot should be consulted to examine the model (Crawley, 2007).

An interesting aspect of this section is that *SD* and *OD* in respect of a given PLSR model can easily be calculated for any unknown observation. Thus, *SD* and *OD* can be derived from spectra of unknown soil samples and be compared to those of the model's cross-validation. Based on this comparison, it is possible to evaluate if a specific PLSR model is suitable for the unknown soil samples (see section 3.3 for an example).

2.5.6. Models based on wavelet coefficients

Wavelet decomposition is a very efficient algorithm for data compressing, e.g. it is used to compress images by the jpg-format. Each observation (spectrum) of a dataset is decomposed into a sum of wavelets that reproduce a unique pattern at different scales and positions. A coefficient is calculated for every scale and position of the wavelet and reflects its contribution to the sum (Lark & Webster, 1999). The wavelet coefficients tell us at which scales, but also at which locations variability in the data is observed.

To understand wavelet decomposition, it may be useful to look at the better known Fourier transforms which decompose the original data into a sum of cosine and sine functions (Alsberg

2. Methods

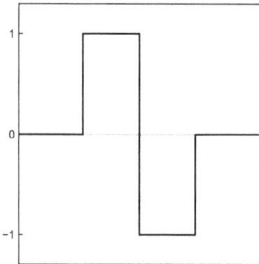 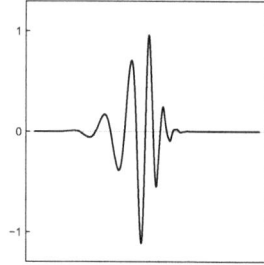

Figure 2.7 The Haar wavelet (left) and Daubechies' so-called extremal phase wavelet with eight vanishing moments.

et al., 1997a):

$$F(w) = \int_{-\infty}^{\infty} x(t) \, e^{-iwt} dt \qquad (2.19)$$

The original data present in the so-called *time domain* are transferred to the *frequency domain*. In the time domain, it is visible **where** variability in the data is present, thus we know the **location**. In the frequency domain, we can see **at what scale or frequency** the data is varying, e.g. if it is fluctuating fast or slowly, but we cannot see where. All information about the location is lost during the Fourier transformation, because periodic functions are used. For wavelet transforms, functions that are different from zero only at a small range are used instead. Consequently, information about the location can be maintained to some degree.

Several families of wavelet functions have been designed. The Haar wavelet (figure 2.7, left side) is the simplest, but it is not a smooth function. More complicated wavelet functions were proposed (among others) by Daubechies (2006), e.g. the so-called extremal phase wavelet with eight vanishing moments used for the calculations within this work (figure 2.7, right side). The *vanishing moments* are a characteristic of each wavelet function: if a function has m vanishing moments, it is able to compress a polynomial of order $m-1$ perfectly (without residuals). Increasing the vanishing moments leads to smoother functions that are more efficient in data compression, but their localisation is reduced (Lark & Webster, 1999). All wavelet function possess two characteristics: they are different from zero at a relatively small range and are zero outside this range, and their integral is zero.

The selected wavelength function is used for all scales and positions of the wavelet transforms: its size is adapted to the different scales, and it is shifted along the x-axis, but the shape remains the same (as demonstrated for the Haar wavelet in figure 2.8). For the coarsest scale 0, the wavelet function covers the whole range of the data series (for spectral data: the whole VNIR spectrum). When going to the next finer level 1, the length of the wavelength is halved. This process is called *dilution*. At this scale, the wavelet can be shifted, because it is narrower now. More complicated wavelets are longer than the Haar wavelet and therefore excess the ends of the data series when shifted to some locations. To avoid problems near the ends of the data series, Cohen et al. (1993) proposed modified wavelet functions which were used for this work.

Within this work, the discrete wavelet transform was applied using the *pyramid algorithm*

2.5. Calibration

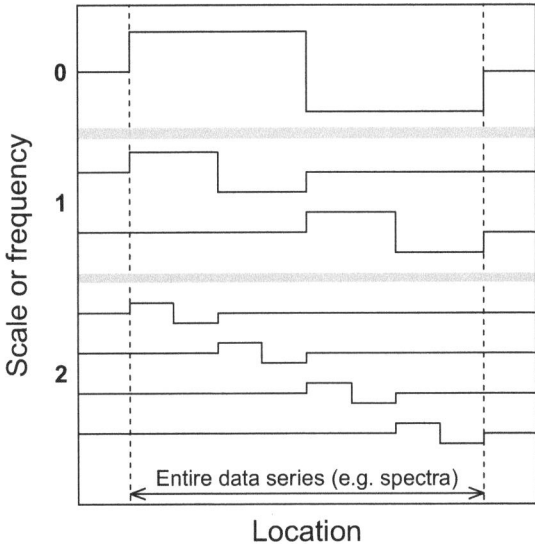

Figure 2.8 Haar wavelets for the scales 0 to 2: the wavelet is halved in size when going to the next finer scale (*dilution*), and the wavelet is shifted along the x-axis at every scale.

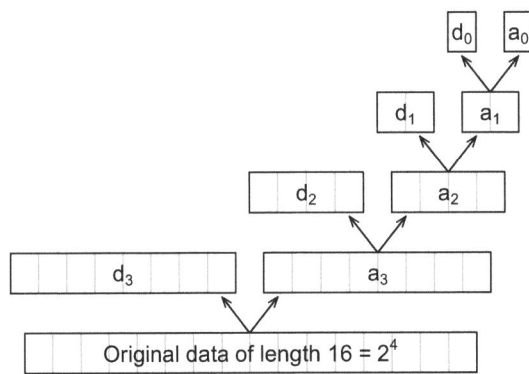

Figure 2.9 Illustration of Mallat's pyramid algorithm for the discrete wavelet transform for a data vector of length $16 = 2^4$. Detail coefficients d_i and approximation coefficients a_i for the scales $i = 3, 2, 1, 0$ are derived in a hierarchical procedure where the length of the data vector is halved by every step.

2. Methods

introduced by Mallat (1989). To suit the algorithm, the length of the data series must be a power of two (2^n). The algorithm is hierarchical and starts at the finest scale which is scale $n-1$. Filters are applied to extract 2^{n-1} so-called *detail coefficients* and 2^{n-1} *approximation coefficients*. The first are the coefficients describing the variability relative to this scale and are usually just called wavelet coefficients. The filters for the next scale $n-2$ are then applied to the 2^{n-1} approximation coefficients to derive 2^{n-2} detail and approximation coefficients each. This procedure is continued until scale 0 where $2^0 = 1$ detail and one approximation coefficient result (figure 2.9). Thus, each data series of length 2^n is transformed into $2^{n-1} + 2^{n-2} + \ldots + 2^1 + 2^0 = 2^n - 1$ detail coefficients used as wavelet coefficients. To summarise: a large number of coefficients representing narrow ranges is calculated for fine scales, whereas few coefficients covering broad ranges are produced for coarse scales.

A large part of the information is present in relatively few wavelet coefficients, whereas the remaining coefficients have small values and are considered to represent mostly noise and unimportant information. Therefore, a small number of selected wavelet coefficients is usually sufficient to reconstruct the original data. By discarding unimportant coefficients, the reconstructed data are compressed and denoised, and could be used for subsequent regression analysis. But wavelet coefficients can also be directly used in linear regressions, because the wavelet coefficients are decorrelated (Viscarra Rossel & Lark, 2009). And there the question arises, how the relevant wavelet coefficients should be selected.

Viscarra Rossel & Lark (2009) proposed to order the wavelet coefficients with respect to their variance (the variation of each wavelet coefficient across all considered spectra) and to include those with highest variance. They argued that variance was a measure of the information present in each wavelet coefficient. The proposed selection criterion as well as other criteria (covariance, Pearson's and Spearman's correlation coefficients, and mean absolute deviation) were tested and compared within this work (chapter 4). The optimum number of wavelet coefficients must be determined by cross-validation, because it depends on the dataset, the used wavelet function, and the regression model used.

The selected wavelet coefficients can be used with any linear or non-linear regression model. For the present work, a quadratic model was chosen:

$$y = \alpha + \beta_1 x_1 + \gamma_1 x_1^2 + \beta_2 x_2 + \gamma_2 x_2^2 \ldots + \beta_n x_n + \gamma_n x_n^2 - d \tag{2.20}$$

where y represents the target variable, d the model's residuals, n the number of used wavelet coefficients, x_1 to x_n the selected wavelet coefficients, and α, β_i and γ_i the model parameters.

At this point, the basic principles of regressions based on wavelet coefficients should be clear; more details can be found consulting the cited references. The tutorials by Lark & Webster (1999) and Alsberg et al. (1997a) as well as the book by Nason (2008, especially for those using R) are recommended to start with. To summarise shortly the approach used for this work (cf. flow chart in figure A.6, page 108): First, the spectra were truncated to contain only $412, 416, \ldots, 2456$ nm wavelengths, so that their length equaled 2^9. Then, wavelet coefficients were calculated from the spectral data using Daubechies' extremal phase wavelet number eight and the modified wavelet functions presented by Cohen et al. (1993). This is illustrated in figure 4.1 (page 50) for a randomly selected VNIR spectrum. From these coefficients, a small number was selected by a specific criterion and used in a quadratic regression model. The resulting models were examined using the diagnostic plots commonly used for linear regression models (cf. Crawley, 2007). Outliers were removed and the model subsequently recalculated.

2.6. Validation statistics

Validation statistics are needed to assess the performance of regression models, to compare different models, and to compare different analytical methods. The first things users want to know when using an analytical method or regression model is: Is it accurate? How big is the error for my estimates? Different sources contribute to the expected error (Hastie et al., 2009):

$$\text{Expected error} = \text{Irreducible error} + \underbrace{\text{Bias}^2 + \text{Variance}}_{\text{MSE}} \qquad (2.21)$$

The irreducible error is caused by the variance of the target variable around its true mean and cannot be avoided. The second and the third terms are related to the used model/method and their sum is represented by the *mean squared error (MSE)*. The bias reflects the mean deviation of the model's estimates from the true mean and is also called *systematic error*. The variance reflects the expected squared deviation of the estimated values around its mean. Usually, the bias is reduced for more complex models, but the variance is increased at the same time. Statisticians denoted this effect as *Bias-Variance trade-off*.

Very often, *root mean squared error (RMSE)* is used instead of MSE ($= \text{RMSE}^2$). It is calculated considering the deviations of values estimated by a specific method or model from the 'true' values (Workman, 2008):

$$\text{RMSE} = \sqrt{\sum_{i=1}^{n} d_i^2 / n} \qquad \text{with } d_i = \hat{y}_i - y_i \qquad (2.22)$$

where $\hat{y}_i$ are the estimated values and y_i the corresponding reference values. Calibration RMSE compares the y-values produced by a model to those used as input, while cross-validation and validation RMSE compare y-values of validation samples with the corresponding reference values. To judge the accuracy of models, cross-validation or validation RMSE must be used because the expected error is underestimated by the calibration RMSE.

Cross-validation is a technique that can be used to simulate the validation of a model when there is now validation dataset available. For k-fold cross-validation, the calibration dataset is divided into k groups (called *segments*). The model is recalculated using $k-1$ segments, and this model is then used to estimate the observations of the segment that was not used for calibration. This procedure is repeated until all segments have been estimated (cf. figure A.8, page 110). Usually, using five to ten segments seems reasonable. It is important to redo all calculations and decisions related to the model during every cross-validation iteration, because the model must not be influenced by the segment used for validation (Hastie et al., 2009). It should also be ensured that identical or similar observations within the dataset fall into the same segment. Otherwise, cross-validation RMSE overestimates the model's accuracy, because some validation samples are (almost) identical to some calibration samples. This seems important in particular for spectral datasets which usually are large and contain various spectra of similar origins. Therefore, when – for example – a dataset contains soil samples originating from various soil profiles, all samples originating from the same profile should fall into the same segment. For the same reason, *leave-out-one* cross-validation, where only one sample is excluded for each iteration, seems not appropriate for most collections of soil spectra. For this work, six-fold cross-validation was used.

Because the RMSE reflects an absolute value and is not scale-invariant, it is sometimes difficult

2. Methods

to compare results calculated for datasets that exhibit very different standard deviations or data ranges. The *coefficient of variation (CV)* puts the RMSE in relation to the mean of the used data (Workman, 2008):

$$CV_{\text{RMSE}} = \frac{\text{RMSE}}{|\bar{y}|} \quad \text{with } \bar{y} = \sum_{i=1}^{n} y_i/n \tag{2.23}$$

As explained above, the bias reflects systematic deviations of the estimated values relative to the true values:

$$\text{Bias} = \sum_{i=1}^{n} d_i/n \quad \text{with } d_i = \hat{y}_i - y_i \tag{2.24}$$

The variance included in the RMSE can easily be calculated by the relation

$$\text{Variance} = \text{RMSE}^2 - (\text{Bias})^2 = \text{MSE} - (\text{Bias})^2 \tag{2.25}$$

which can be derived from equation 2.21.

The *reproducibility* of analytical methods is assessed by comparing the analyses of replicate samples. The *standard error of laboratory (SEL)* quantifies the deviations between r replicate analyses of n samples (Workman, 2008):

$$\text{SEL}^2 = \sum_{j=1}^{r} \sum_{i=1}^{n} (y_{i,j} - \bar{y}_i)^2 / [n(r-1)] \quad \text{with } \bar{y}_i = \sum_{j=1}^{r} y_{i,j}/r \tag{2.26}$$

For duplicate analyses, equation 2.26 can be simplified to

$$\text{SEL}^2 = \sum_{i=1}^{n} (y_{i,1} - y_{i,2})^2 / (2n) \tag{2.27}$$

with $y_{i,1}$ and $y_{i,2}$ being the first and second analyses, respectively. A CV_{SEL} can be defined analogically to CV_{RMSE} (equation 2.23).

In literature, regression models are often assessed using the coefficient of determination R^2. For linear regression models, it is defined as (Crawley, 2007):

$$R^2 = 1 - \frac{\text{Error sum of squares}}{\text{Total sum of squares}} = \frac{\text{Explained variance}}{\text{Total variance}} \tag{2.28}$$

Good regression models explaining most of the dataset's variability achieve R^2 near one, whereas models explaining little variability have low R^2. To judge the accuracy of models, R^2 is of little use because – as visible from its definition – it is influenced by both the total variance and the explained variance. Furthermore, RMSE has all the requested information. Thus, R^2 should only be reported supplementary to other statistics like RMSE. Unfortunately, R^2 is very popular because it is considered to be easily interpretable and understandable (like *p*-values), and it is used in many published articles without reporting RMSE. Due to the mentioned reasons, R^2 was sparsely used throughout this work.

2.7. Software: Using R

R is a free software environment for statistical computing and graphics distributed under the terms of the GNU General Public License (R Development Core Team, 2011). All calculations and most graphics of the present work were conducted using R version 2.13.0. In addition to the basic packages provided by the program, the package wavethresh for wavelet decomposition (Nason, 2008), and numerous routines written by me to cover my specific needs for spectral data processing, calibration, and validation were used. Introductions into R are provided by the project's homepage (www.r-project.org) and by Crawley (2007).

3. VNIRS laboratory models

Within the two first sections of this chapter, models to estimate organic and total carbon (C_{org} and C_{tot}), organic matter (OM), and total nitrogen (N_{tot}) from VNIRS spectra will be derived and cross-validated using the methodology presented in chapter 2. Only PLSR was considered for this chapter, calibrations using wavelet transforms will be discussed in the following chapter. In section 3.3, derived models to estimate C_{org} will be validated by soil samples originating from a distinct area. And finally, reproducibility and uncertainties of VNIRS estimates will be addressed in section 3.4.

3.1. Estimating soil carbon

Organic carbon is the main constituent of soil organic matter which comprehends the living biomass, dead biomass at different stadiums of decay, and humus representing stable decomposition products of biomass. Soil quality is strongly influenced by the content of humus due to its capacity to withhold water and nutrients, due to its ability to promote aggregation and biological activity, and due to its ability to absorb heavy metals and organic pollutants (Gisi et al., 1997). Soil organic content is one of the most important parameters to assess and monitor soil quality. Beside organic carbon, total carbon additionally includes inorganic carbon, in soils mostly present as carbonate minerals. Due to its geological origin or due to weathering, inorganic carbon is absent in some soils (Nelson & Sommers, 1996).

Within the present work, models to estimate OM, C_{org} and C_{tot} from spectral data were derived from various ensembles of soil samples. This section provides an overview of the performance of models derived by PLSR. To facilitate the comparison of the derived models – and because OM according to the Swiss standard method (FAL, 1996) is determined by measuring C_{org} and multiplying the result by 1.725 –, OM contents were transformed back to C_{org} contents.

3.1.1. Soil samples and methods

The first set of soil samples was designated *nabo-eic* and consisted of samples provided by the Swiss Soil Monitoring Network (NABO), Zürich-Reckenholz, and the University of Applied Sciences Changins (École d'Ingénieurs de Changins, EIC), Nyon. The NABO samples originated from 105 sampling sites covering whole Switzerland including a large variety of soils. Depending on the availability, one to four replicate samples were measured per sampling site. The EIC samples originated from a collection of soils used for educational purposes and represented a diverse set of soils predominantly from alpine soils, most of them with low carbon contents. All samples were measured for C_{org} by dichromate digestion according to FAL (1996). Soil samples with C_{org} contents above $120\,g\,kg^{-1}$ were excluded because there were only few samples in that range. A second set of soil samples, *nabo-eic <60*, was derived by using only samples with C_{org} contents below $60\,g\,kg^{-1}$ from the *nabo-eic* set.

Another set of soil samples, *fr-so*, included samples originating from 469 sites in north-western Switzerland (Cantons of Fribourg and Solothurn) provided by the Soil Protection Unit of the Canton

3. VNIRS laboratory models

Table 3.1 Studied sets of soil samples to estimate carbon contents. n_{total}: number of samples in the original set; $n_{s.out}$: number of spectral outliers (removed by pre-screening procedure); $n_{m.out}$: number of model outliers (removed during calibration step); n_{used}: number of samples used for calibration and validation; sd: standard deviation; statistics refer to the used soil samples.

Dataset	n_{total}	$n_{s.out}$	$n_{m.out}$	n_{used}	Min.	Median	Mean	Max.	sd
						— (g C kg^{-1}) —			
nabo-eic	427	84	1	342	3	22	25	92	16
nabo-eic <60	400	119	4	227	3	20	22	56	13
fr-so	789	79	10	700	4	15	17	42	7
rosslau	243	49	8	186	3	18	18	39	8
lu	64	2	0	63	17	31	32	49	7

of Solothurn and the Soil Monitoring Unit of the Canton of Fribourg (FRIBO). Approximately 80 % of these had been taken as mixed samples from the top soil layer (0-20 cm), and the remaining had been collected by horizons within the 0 to 30 cm depth layer. The available OM data were determined according to FAL (1996). These were transformed back to C_{org} contents. Sixteen soil samples with C_{org} contents bigger than 60 g kg^{-1} were removed, because very few samples were available for this range. The *fr-so* set represented a regional dataset consisting predominatly of samples from the Swiss Central Plateau ('Mittelland'). This region covers about 30 % of Switzerland and is located between the Alps and the Jura mountains; it is situated between 400 and 700 m altitude, it is densely populated and the country's most important agricultural and industrial region. The *fr-so* set was under no circumstances representative for the Central Plateau, but the central to western parts were reasonably covered.

The *rosslau* set consisted of soil samples collected from a test site near Rosslau, Saxony-Anhalt, Germany, used by the iSOIL project. The site is situated close to the river Elbe and parts of it are regularly flooded by the river. Three soil layers (0-10, 10-30, and 30-70 cm) were sampled at 81 sampling sites. For this set, C_{tot} contents were determined by dry combustion. The test site included few ares, thus *rosslau* represented a local dataset.

A second local set of soil samples, *lu*, consisting of 64 samples collected from agricultural plots in Emmenbrücke, Canton of Lucerne, Switzerland, was used. The set included mixed samples from the top soil layer (0-20 cm) as well as single samples from three fractions of the topsoil (0-5, 5-10, and 10-20 cm). None of the samples contained inorganic carbon, therefore the C_{tot} contents determined by dry combustion were considered equal to C_{org}. As a specific characteristic, two spectra had been collected from every sample (and were used together in the models): one with the FS Pro device, and one with the FS 3. All previous sets were measured either only by the FS Pro (*nabo-eic, fr-so*), or by the FS 3 (*rosslau*).

The reference data for the *lu* set were determined by the University of Bern (Thanks to Isabel!), those of the *rosslau* set by our iSOIL project partners, and the remaining by the owners of the respective soil archives. The statistics of the used sets of soil samples are provided by table 3.1.

All spectral data were collected from dried and sieved soil samples using an ASD FieldSpec Pro or a FieldSpec 3 Hi-Res (ASD Inc., Boulder, CO, USA) with a muglight contact probe and a Spectralon panel for white reference. The measuring procedure described in section 2.3 and

3.1. Estimating soil carbon

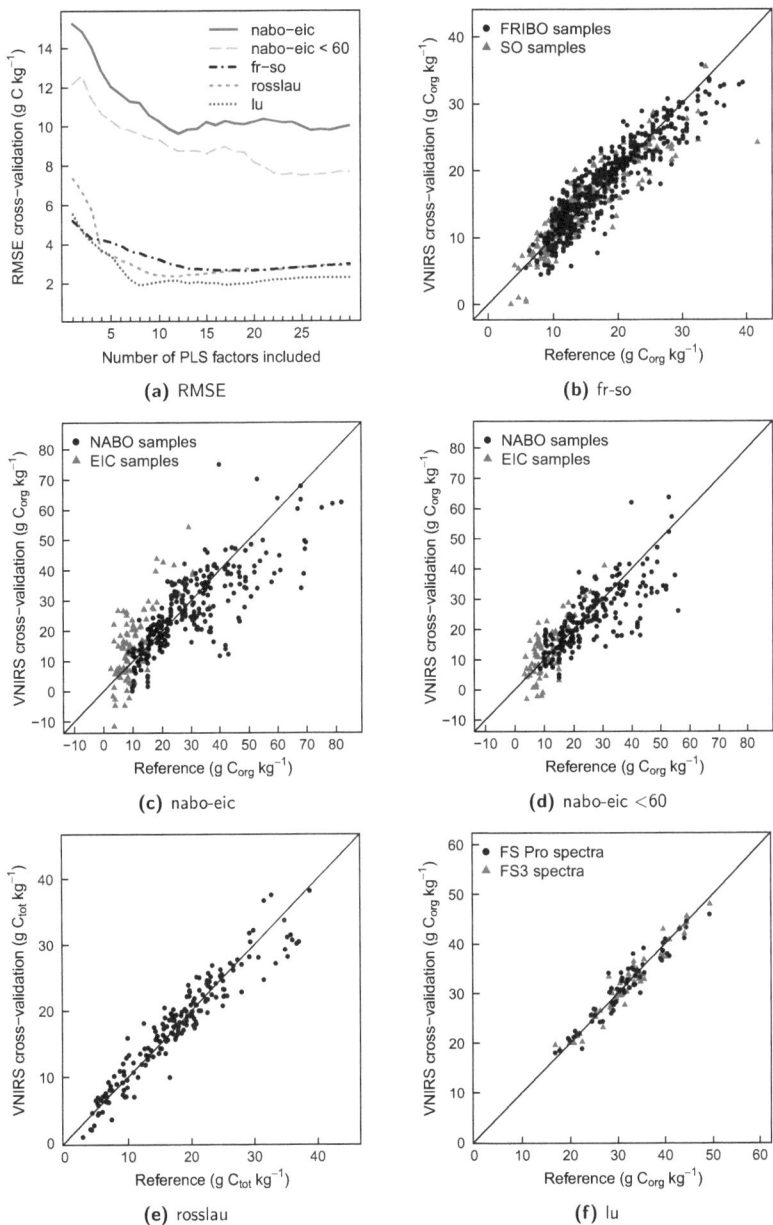

Figure 3.1 PLSR models to estimate organic and total carbon: RMSE and model estimates.

3. VNIRS laboratory models

Table 3.2 Performance of PLSR models (cross-validation RMSE) to estimate C_{org} and C_{tot}.

Dataset	Extension	Target variable	RMSE ($g\,C\,kg^{-1}$)	Used factors
nabo-eic	national	C_{org}	9.8	13
nabo-eic <60	national	C_{org}	7.6	22
fr-so	regional	C_{org}	2.7	13
rosslau	local	C_{tot}	2.4	10
lu	local	$C_{org} = C_{tot}$	1.9	8

figure A.2 (page 104) was used. The spectral data were transformed to absorbance. For each set, models were calibrated using PLSR according to the procedure explained in section 2.5 and figure A.4. Their performance was assessed by six-fold cross-validation.

3.1.2. Results

For the local and regional sets of soil samples, good cross-validation RMSE between 2 and $3\,g\,C\,kg^{-1}$ were achieved (cf. table 3.2), whereas very high RMSE between 7.5 and $10\,g\,C\,kg^{-1}$ were observed for the models of both *nabo-eic* sets. For the latter, the cross-validation estimates strongly scattered (figure 3.1). For the local and regional sets, RMSE decreased smoothly with increasing numbers of PLS factors and reached its minimum roughly near ten factors. The corresponding curves of the *nabo-eic* models were uneven and they required more PLS factors to reach their minimum.

To assess the importance of the single wavelengths for the different models, regression coefficients derived by the models were compared (figure 3.3). The coefficients derived from the local datasets (*rosslau*, *lu*) resembled strongly. Important wavelengths to predict carbon contents were identified in the visible (350-760 nm), around 1400 and 1900 nm as well as above 2100 nm with very prominent features near 2200 and 2300 nm. For the model based on the *fr-so* set, important wavelengths seemed to be more evenly distributed over the whole spectral range. The features near 1400, 1900, 2200, and 2300 nm were also observed for this model. While the coefficients derived for *nabo-eic* resembled roughly those derived for *fr-so*, those derived for *nabo-eic <60* looked clearly distinct and were fluctuating strongly.

3.1.3. Discussion

Organic and total carbon are very popular soil properties to be analysed with VNIRS and numerous studies were published. A comprehensive overview of them is given by Stenberg et al. (2010). The authors observed a strong correlation of the reported RMSE of C_{org} estimation models to the standard deviation of C_{org} within the used dataset: calibrations based on datasets exhibiting a higher variation with respect to C_{org} also exhibited higher RMSE. The RMSE of the models derived from the local (*rosslau*, *lu*) and regional (*fr-so*) datasets of the present work were comparable to those of the studies considered by the overview study (figure 3.2). In contrast, the models derived for the *nabo-eic* datasets performed significantly worse than could be expected from their C_{org} standard deviation. Either these datasets were corrupted – e.g. a lot of bad spectral measurements

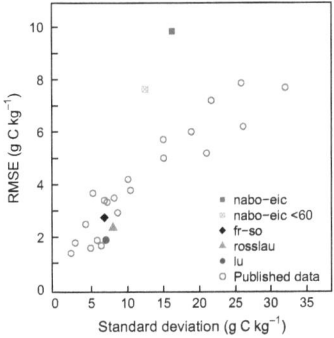

Figure 3.2 RMSE and standard deviation of models to estimate total and organic carbon compared to those of published works (reproduced and modified from Stenberg et al., 2010).

due to problems with the white reference panel –, or the bad performance was induced by some characteristics of these datasets.

As no measuring errors became evident, it was assumed that the characteristics of the datasets caused the differences in performance. The studied datasets varied strongly in their geographical extension from local to regional to national. While the soils included in local sets usually were rather homogeneous, those of non-local sets were much more diverse and usually the variation within C_{org} contents was larger too. Among the datasets of the present work, the regional *fr-so* set exhibited a C_{org} standard deviation comparable to the local datasets, but its soils were very heterogeneous compared to them. Indeed, a slightly higher RMSE was observed for the *fr-so* model compared to the local models. It was suggested that, besides from the C_{org} standard deviation, also the heterogeneity of the included soil samples influenced the resulting RMSE. Or possibly, soil diversity might have been the main factor affecting RMSE, but as usually variation in C_{org} and soil diversity are correlated, C_{org} standard deviation suited to explain most of the differences in RMSE.

The *nabo-eic* datasets were more diverse than the regional *fr-so* set. They also included a lot of 'extreme' soils, e.g. originating from alpine sites. To visualise the heterogeneity of the used sets of soil samples, the absorbance spectra of all datasets were pooled in one dataset to conduct a principle component analysis. The scores of the first principle components are displayed in figures 3.4 and A.9. The points corresponding to *nabo-eic* were much more scattered than those of the remaining datasets. In addition, there is a lot of empty space between single points of *nabo-eic*. Presumably, the database of the *nabo-eic* models was to weak with respect to the captured soil diversity. It was remarkable that, although being much more diverse, the *nabo-eic* sets consisted roughly of half as many samples as the *fr-so* set. To cover reasonably the soil diversity observed in the *nabo-eic* sets, a much bigger number of soil samples would have been needed. The scattering within the principle components space could be used as proxy for the diversity of the soil samples. I would suggest that the covered space should be evenly and densely populated by data points (as observed for the other studied datasets) to enable successful calibrations by PLSR.

The comparison of the regression coefficients (figure 3.3) revealed very similar regression coefficients for the models derived from the two local sets of soil samples (*rosslau*, *lu*). The resemblance of these two models was striking, but also the models derived for *fr-so* and *nabo-eic* exhibited many similarities with the models of the local datasets. Only the regression coefficients derived for *nabo-eic <60* looked very different, possibly a consequence of the large number of PLS factors included. The observed importance of the visible range to estimate soil carbon did not

3. VNIRS laboratory models

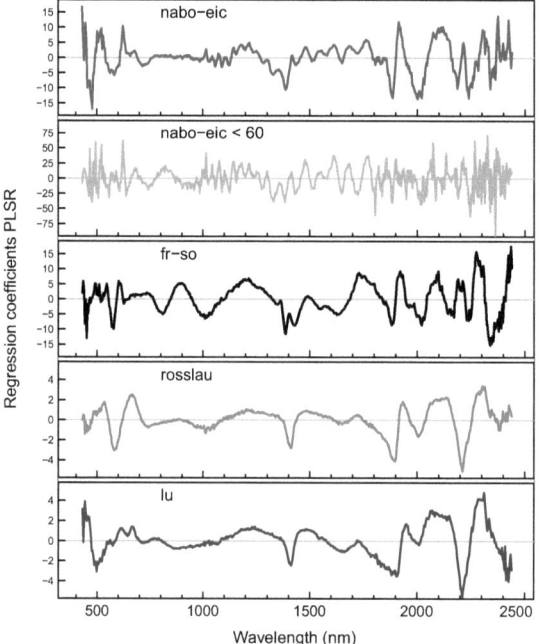

Figure 3.3 Regression coefficients of the derived PLSR models to estimate organic and total carbon. (Positive regression coefficients imply that samples with increased carbon contents generally exhibited increased absorbance at the corresponding wavelengths, whereas negative coefficients imply decreased absorbance for increased carbon contents.)

surprise because it has been well known for a long time that soil colour is strongly linked to soil carbon. The importance of the bands near 1400 and 1900 nm were attributed to reported absorption features of carboxylic acids (overtones of C=O fundamental at 1449 and 1930 nm; Viscarra Rossel & Behrens, 2010) as well as to absorptions of hydroxy groups (at 1400 nm, when interacting with water also near 1900 nm; cf. table 2.2) which are numerous in organic matter. Different organic compounds could have been responsible for the observed features above 2100 nm, e.g. aliphatics (2275 nm), methyls (above 2300 nm), or polysaccharides (2137 nm; Viscarra Rossel & Behrens, 2010).

The locations of the wavelengths important for estimating soil carbon near 1400 and 1900 nm coincided with the absorption features of water molecules and hydroxy groups observed for spectra of dried soil samples. It was shown that these spectral regions as well as wavelengths above 2000 nm are particularly sensitive to changes in soil moisture (c.f. chapter 5). The absorbance of these wavelengths is easily altered by small variations in soil moisture, e.g. caused by changes in air humidity. Thus, in order to derive regression models that are more stable on the long-term and for larger collections of soil samples, it would be favourable making these spectral regions less influential. As the regression coefficients for soil carbon exhibited features over the whole VNIR range, it seemed very probable that soil carbon could be estimated without using the water-sensitive ranges. In section 6.3.2, an example will be shown where discarding these ranges indeed improved the stability of the resulting model substantially without degrading its accuracy. Of course, the effect of the described wavelength removal usually cannot be detected in cross-validation, because

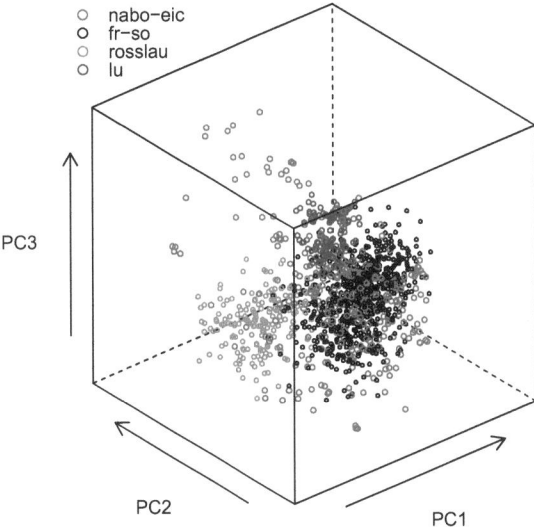

Figure 3.4 Three-dimensional scatter plot of the first three principle components of the pooled dataset of all absorbance spectra used for the different PLSR models to estimate total and organic carbon. (See figure A.9, page 112, for the first five principle components of the same dataset.)

often the samples used for calibration originate from the same collection and were measured by VNIRS within the same campaign.

The comparison of the regression coefficients for the different models also depicted very clearly the disadvantages of using too many PLS factors in regression models. It was evident that including more factors induced larger regression coefficients. The model derived for the *nabo-eic <60* dataset included 22 factors and its regression coefficients were roughly five times bigger than those derived for the *nabo-eic* and *fr-so* sets including 13 factors. What is the problem about the larger regression coefficients? They imply that small differences in the spectra (due to real differences of the soil sample or due to measuring errors) have a much bigger impact on the results. Therefore, models exhibiting bigger coefficients are less stable and more prone to errors. Usually, this effect is denoted over-fitting. To achieve an optimum performance on the long-term and beyond the calibration samples, it is essential to use parsimonious models. Additional PLS factors should be only included if a substantial RMSE reduction results. Under this aspect, the model exhibiting the lowest cross-validation RMSE seems not to be best in every case.

3.2. Estimating total nitrogen

Nitrogen is the most important macro-nutrient required by plants. On average, over 99 % of the soil's nitrogen is present in organic forms: this fraction of nitrogen is incorporated in organic matter and its biological availability strongly varies according to the molecular structures of the present organic matter. The concentrations of the plant-available inorganic forms, predominantly ammonium (NH_4^+) and nitrate (NO_3^-), are highly variable in time and location. Large amounts of nitrogen are applied to agricultural fields as organic or mineral fertilsers. Inadequate soil management and/or anaerobic soil conditions can result in considerable losses of nitrogen to aquatic systems and the athmosphere (Gisi et al., 1997).

3. VNIRS laboratory models

Table 3.3 Studied sets of soil samples to estimate nitrogen contents. n_{total}: number of samples in the original set; $n_{s.out}$: number of spectral outliers (removed by pre-screening procedure); $n_{m.out}$: number of model outliers (removed during calibration step); n_{used}: number of samples used for calibration and validation; sd: standard deviation; statistics refer to the used soil samples.

Dataset	n_{total}	$n_{s.out}$	$n_{m.out}$	n_{used}	Min.	Median	Mean	Max.	sd
						— (g N kg^{-1}) —			
wsl	653	86	17	550	0.0	0.8	1.4	9.5	1.7
rosslau	243	49	8	186	0.2	1.8	1.8	4.0	0.8
lu	65	2	0	63	1.9	3.3	3.4	5.1	0.8

3.2.1. Soil samples and methods

Three different sets of soil samples were used to derive models to estimate N$_{tot}$ by PLSR: two sets of local extension (*rosslau*, *lu*), and one set covering whole Switzerland (*wsl*). The local sets of soil samples were previously used to estimate soil carbon and were described in section 3.1.1. The *wsl* set consisted of 567 soil samples originating from 115 forest sites spread across all regions and landscapes of Switzerland and were provided by the Swiss Federal Institute for Forest, Snow and Landscape Research (WSL), Birmensdorf, Switzerland. The soil samples had been collected by horizons from the whole profiles. The analytical data for N$_{tot}$ (determined by dry combustion) were provided by the owner of the archive. N$_{tot}$ contents of the local datasets were determined by dry combustion by our iSOIL project partners and the University of Bern, respectively. Table 3.3 provides an overview of the used sets of soil samples. Spectral data and PLSR models were conducted according to the procedure described for the models to estimate soil carbon (cf. section 3.1.1).

3.2.2. Results

For the soil samples of the *wsl* set, a cross-validation RMSE near 0.6 g N kg^{-1} was achieved. For the local datasets, *rosslau* and *lu*, clearly lower RMSE near 0.2 g N kg^{-1} resulted (table 3.4). The scatter plots of the six-fold cross-validation of these models resembled those of the corresponding models to estimate soil carbon (figures 3.5 and 3.1). Similarly, the decrease in cross-validation RMSE with increasing numbers of used PLS factors was smooth and reached its minimum near ten factors. In contrast, the model derived for the *wsl* set exhibited clearly higher RMSE, and the curve representing RMSE was rather uneven. Estimates by this model of samples exhibiting low nitrogen contents (< 1 g kg^{-1}) as well as those above 5 g kg^{-1} seemed to be problematic.

The comparison of the regression coefficients of the models derived from the three datasets revealed a lot of similarities (figure 3.6). The coefficients derived for the local datasets resembled strongly those of the corresponding models to estimate soil carbon (figure 3.3). For all N$_{tot}$ models, important features within the regression coefficients were observed near 1400, 1900, 2000, and between 2100 and 2200 nm. The magnitude of the regression coefficients was the same for all models.

3.2. Estimating total nitrogen

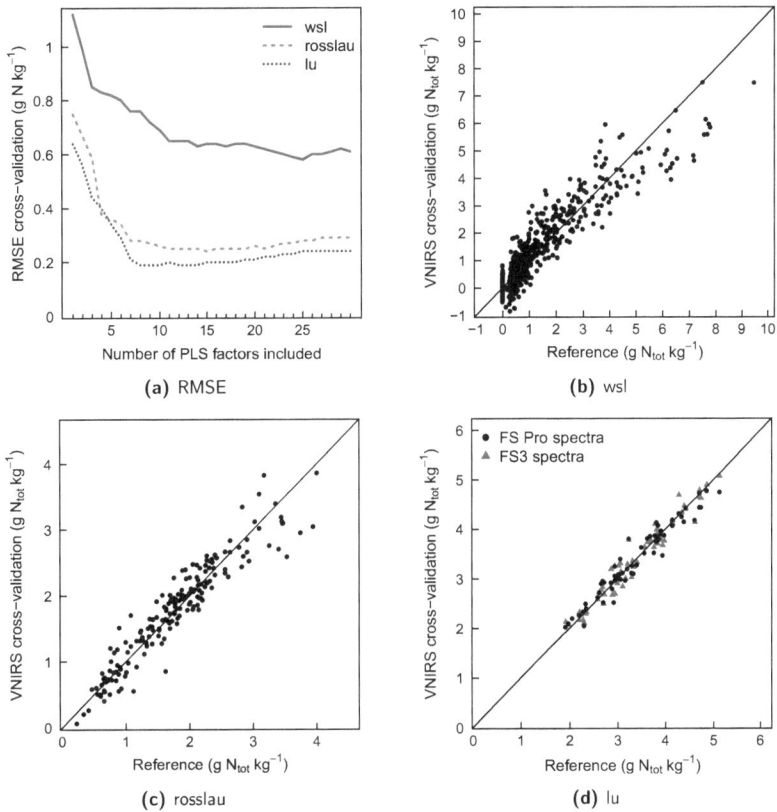

Figure 3.5 PLSR models to estimate total nitrogen: RMSE and model estimates.

Table 3.4 Performance of PLSR models (cross-validation RMSE) to estimate N_{tot}.

Dataset	Extension	Target variable	RMSE (g N kg^{-1})	Used factors
wsl	national	N_{tot}	0.63	14
rosslau	local	N_{tot}	0.24	10
lu	local	N_{tot}	0.19	8

3. VNIRS laboratory models

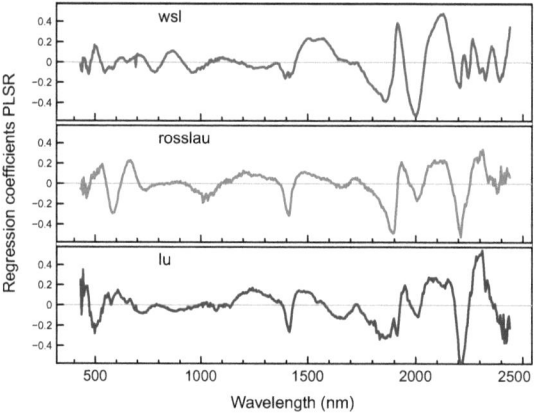

Figure 3.6 Regression coefficients of the derived PLSR models to estimate total nitrogen. (Positive regression coefficients imply that samples with increased nitrogen contents generally exhibited increased absorbance at the corresponding wavelengths, whereas negative coefficients imply decreased absorbance for increased nitrogen contents.)

3.2.3. Discussion

Using NIRS, Chang et al. (2001) were able to estimate N_{tot} of soil samples from the top 30 cm with a RMSE of 0.62 g kg^{-1} by selecting spectrally similar calibration samples and applying principle component regression. Comparable cross-validation RMSE between 0.5 and 0.6 g N kg^{-1} were reported by Fystro (2002) for 80 top and sub soil samples originating from Norwegian grassland fields. Also using NIRS and PLSR, Brunet et al. (2007) achieved cross-validation RMSE between 0.20 and 0.26 g kg^{-1} for < 2-mm sieved samples.

RMSE reported for the three models derived in the present work conformed to those reported by the studies mentioned above. The two local datasets exhibited RMSE comparable to those reported by Brunet et al. (2007) who also used a dataset of local extension (samples from eight tropical sites, mean 0.6 g N kg^{-1}, maximum 2.3 g N kg^{-1}). The larger dataset *wsl* covering whole Switzerland achieved a RMSE comparable to those reported by Chang et al. (2001) and Fystro (2002). The observed differences between local sets of soil samples and sets of greater extension did not surprise, because it seemed reasonable that – as for C_{org} models – datasets with larger variation in N_{tot} resulted in elevated RMSE. The low RMSE of the local datasets of this work were notable because these exhibited clearly higher N_{tot} contents compared to Brunet et al. (2007). Usually, higher values in the estimated variable induce elevated RMSE.

The *wsl* set of soil samples shared most characteristics with the *nabo-eic* sets used to estimate C_{org} contents: national extension, including a large diversity of soils, including 'extreme' soils, and a rather limited number of soil samples compared to the diversity of included soils. The uneven curve visible for the RMSE in dependency of the number of included PLS factors led to the conclusion that the derived model for the *wsl* set might be unstable. Nevertheless, the achieved RMSE comparable to those of published studies and the clear features visible in the plot of the regression coefficients (figure 3.6) indicated that the model was not too bad. Although the number of soil samples was rather low with respect to the captured soil diversity, it was still clearly higher as for the *nabo-eic* models. Anyhow, it was recommended to enlarge the database in order to establish a calibration for N_{tot} of national extension.

The inspection of the regression coefficients of the N_{tot} models revealed that spectral ranges important to estimate N_{tot} contents were scattered over the whole VNIR range (figure 3.6). Various absorption features were reported for organic compounds containing nitrogen, e.g. at 751, 1000, 1500, and 2060 nm for N-H bonds in amines and at 1524 and 2033 nm for C=O in amides (Viscarra Rossel & Behrens, 2010). For ammonia, one of the dominant forms of inorganic nitrogen, three strong absorption bands near 1534, 2000, and 2210 nm were observed when dissolved in water (Workman, 2008). The position of these may be altered by the soil matrix. For urea, additional absorption bands were reported between 1500 and 2300 nm wavelengths (Workman, 2008). No significant absorption can be expected for nitrate (NO_3^-) because it exhibits no N-H bonds.

Commonly, it is believed that N_{tot} is estimated by VNIRS not due to its direct spectral response, but due to correlations with organic carbon. Indeed, both local datasets, *rosslau* and *lu*, exhibited very strong correlations between C_{tot} and N_{tot} (Pearson's correlation coefficients: 0.99 and 0.96, respectively; cf. scatter plots in figure A.11, page 113). Furthermore, the regression coefficients for the respective models derived to estimate C_{tot} and N_{tot} were very similar. The correlation between C_{tot} and N_{tot} coefficients was 0.99 for the *rosslau* and 0.96 for the *lu* set – equal to the correlation of the two soil parameters. Only slight deviations were observed for the regression coefficients, the most remarkable of them near 1900 nm for the *rosslau* models and between 500 and 700 nm, between 1800 and 1950 nm, and near 2100 nm for the *lu* models. The variance explained by the models was also assessed for these models: the N_{tot} models for *rosslau* and *lu* explained 92 and 96 %, respectively, of the total variance of the target variable, while the carbon models explained 93 and 95 %, respectively.

It was assumed that, whenever such strong correlations between soil carbon and nitrogen occur, it cannot be avoided that absorptions of soil carbon are used by N_{tot} models (and vice versa C_{org} models are using N_{tot} absorptions, too). And there are even absorption features referring to both carbon and nitrogen, e.g. the absorption of C=O in amides. The described correlation effects surely also affected the models derived in this work. But assuming that N_{tot} was only estimated by the correlation to soil carbon would imply that N_{tot} models were slightly less accurate than the carbon models. As this was not case, it was concluded that there were at least some nitrogen specific spectral features used by the models. This was supported by the slight deviations between the correlation coefficients as well as the fact that the explained variance of carbon and nitrogen models were similar.

3.3. Applying existing calibrations

The main purpose of calibrating models is to estimate soil parameters, especially from unknown samples. As long as the samples originate from the same ensemble of soil samples as those used for calibration, the estimation errors usually are comparable to those observed in cross-validation (if cross-validation is conducted correctly). The application of a model to a unknown set of soil samples will be assessed within this section. The soil samples of the *lu* set described in section 3.1 were estimated by the models derived from the *fr-so* dataset. Both sets consisted of soil samples originating from the Swiss Central Plateau ('Mittelland'), but from distinct, separated regions. The term 'validation' will be used for the applied procedure knowing that a proper validation would require a more diverse dataset representing the Central Plateau.

3. VNIRS laboratory models

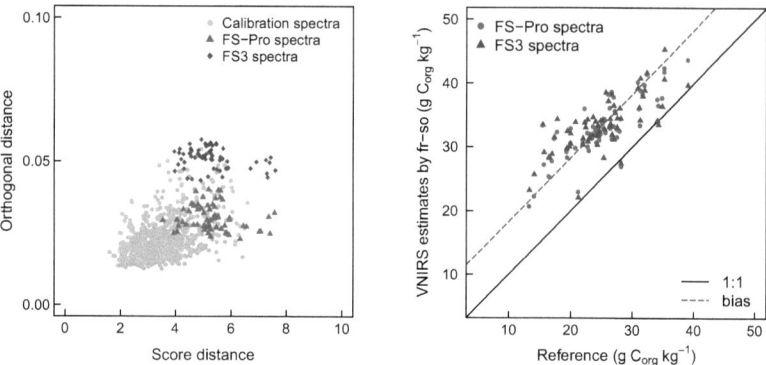

Figure 3.7 Score and orthogonal distances (left) and OM estimates (right) of the *lu* samples by the *fr-so* model including 13 PLS factors.

3.3.1. Methods

C_{org} models derived from the *fr-so* dataset presented in section 3.1 were used to estimate C_{org} contents of soil samples of the *lu* set presented in the same section. The model including the optimum number as determined by cross-validation of 13 PLS factors was used. In addition, models including one to 20 PLS factors were used to assess the sensibility of the validation performance in respect of the number of used factors. Because C_{org} contents of the *lu* samples were determined by dry combustion, while the *fr-so* samples were analysed by the dichromate digestion method according to FAL (1996), the first were multiplied by 0.79 according to the explanations presented in section 2.4.2. The validation performance for the considered models was assessed by comparing RMSE, bias, variance, and scatter plots. In addition, score distances and orthogonal distances of the validation samples in respect of the calibration samples were compared.

3.3.2. Results

The inspection of score and orthogonal distances for the model including 13 PLS factors (the optimum number of factors determined by cross-validation) indicated that the spectra collected by the FS Pro device exhibited distances comparable to those of the calibration samples, while the spectra collected by the FS3 exhibited similar score distances but clearly higher orthogonal distances (figure 3.7). The C_{org} estimates by the model including 13 factors were strongly biased (8.3 g C kg^{-1}), thus a high RMSE of 9.1 g C kg^{-1} resulted. Only slight differences were observed between corresponding C_{org} estimates of the two spectrometers.

The comparison of all models including one to 20 PLS factors revealed that RMSE strongly varied in dependency of the number of included factors (figure 3.8). The variations in RMSE were mainly caused by fluctuations of the bias, while the variance remained relatively stable. For all models including seven to 13 factors, the square root of the variance remained between 3 and 4 g C kg^{-1} which is clearly smaller than the standard deviation of the estimated variable (5.6 g C kg^{-1}). The lowest validation RMSE was achieved by using only two PLS factors, but

3.3. Applying existing calibrations

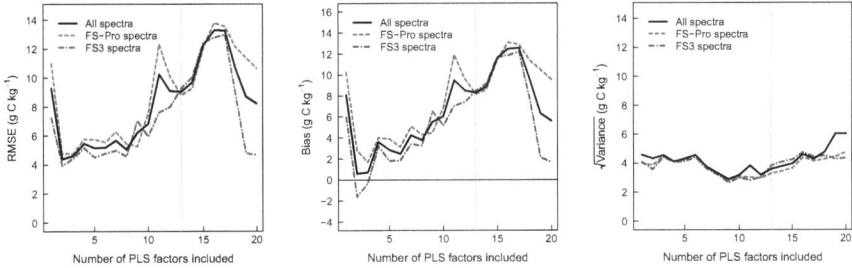

Figure 3.8 RMSE, bias, and variance in respect of the included number of PLS factors for C_{org} estimates of the *lu* samples by *fr-so* models.

RMSE remained rather low for all models from two to nine factors. For all models using more factors, RMSE was considerably higher. The lowest variance was achieved by the model including nine factors. Only small performance differences were observed between the two spectrometers.

3.3.3. Discussion

The presented results indicated that the models based on the *fr-so* dataset can only by used under reservations to estimate soil samples from other regions of the Swiss Central Plateau, even if they exhibit comparable score and orthogonal distances. The estimates for the *lu* samples were strongly biased, and most of the reported RMSE was attributable to the bias term. The rather large contribution of the bias was explained by the fact that a very homogeneous set of soil samples was estimated by models derived from a heterogeneous set of samples. Therefore, the deviation of the model's estimates for the *lu* samples from the reference measurements were expected to be similar for all of them. In other words: if sample *i* was clearly over-estimated by the model, the neighboured samples were expected to be over-estimated as well because they presumably exhibited similar spectral features.

It is possible to take advantage of this effect: if – as in the presented example – a set of similar samples should be estimated by an existing model, the accuracy can be increased substantially by correcting for the bias. If the bias was known exactly, the remaining RMSE would be equal to the square root of the variance (according to equation 2.25). Of course, in reality it is impossible to know the bias exactly, because it must be estimated from few reference analyses. But nevertheless, even by correcting with the estimated bias, the accuracy is still increased and the observed RMSE then equals the square root of Variance + Remaining bias^2. Applying this procedure, an existing model also could be adapted for various homogeneous sub-groups of soil samples.

The results also revealed that the optimum number of PLS factors determined by cross-validation is not necessarily the optimum number to estimate samples of another set of unknown soil samples. For the *lu* samples, the bias was smallest when using only two PLS factors, while the variance was smallest when using nine factors. When comparing the scatter plots of the C_{org} estimates for the models including increasing numbers of PLS factors (figure A.13, page 115), it became evident that the model exhibiting the smallest RMSE accounted poorly for C_{org} variations within the validation set and the low RMSE resulted mainly due to the low bias. In contrast, including

3. VNIRS laboratory models

nine PLS factors was regarded optimal because the variance was lowest and the bias not as high as when using more factors. The rather small changes in variance between seven and 13 included factors indicated that any of these models could be used when bias-correction was applied.

It was reasoned whether the optimum number of PLS factors with respect to the validation set could be derived without using any reference analyses. For this purpose, the score and orthogonal distances for different numbers of included PLS factors were assessed (figure A.12, page 114). It was implied by their definitions that generally score distances are increasing and orthogonal distances are decreasing with increasing numbers of PLS factors included in the model. For small numbers of PLS factors, the validation samples exhibited bigger orthogonal distances than the cross-validation samples, but similar score distances. In contrast, including a large number of PLS factors resulted in orthogonal distances comparable for validation and cross-validation, but bigger score distances for the validation samples. It was supposed that there was a trade-off between minimising score and orthogonal distances and that the best estimations were achieved by selecting a number of PLS factors where both were rather small. And indeed, the cloud of points representing score and orthogonal distances seemed to be located most compactly when using nine factors which corresponded to the model considered best. Therefore, it was concluded that the used distance measures could become useful tools to assess the applicability of existing calibrations on unknown soil samples and that they suited to optimise the used calibration for specific sets of samples. Certainly, further research is needed to assess the behaviour of the used distance measures in other situations and make them better comprehensible, e.g. by some sort of statistic.

Interestingly, only small differences were observed between the estimates derived from the spectra collected by a FS Pro spectrometer compared to those by a FS3. In contrast, the inspection of score and orthogonal distances revealed that the spectra exhibited clear differences, as the spectra of the two devices formed two separate clusters. Because the two clusters mainly differed with respect to the orthogonal distances while the score distances were comparable, it was concluded that the spectral differences were not relevant to the C_{org} model. In other words: the differences disappeared during the data compression conducted by the PLSR algorithm. Furthermore, it was concluded that elevated orthogonal distances do not necessarily result in degraded estimations (e.g. see the plots for $a = 13$ in figures A.12 and A.13). Elevated orthogonal distances imply that more spectral information has been lost during the compression of the data, but as long as the lost information is not relevant to the target variable, no effect is observed.

Comparing the C_{org} estimates calculated from the spectra of the first series measured by a FS Pro device with the corresponding estimates of the second series measured several months later by a FS 3 device revealed that the bias between the two series as well as the variance of the replicate analyses strongly depended on the number of PLS factors included in the model (figure A.14). Several effects might have contributed to the observed differences: differences between the spectrometers, alteration of the soil samples (mainly due to changes in air humidity), and differences between the measured sub-samples due to soil variability within the samples. To assess more exactly the contributions of these parameters, especially the one of differences between the spectrometers, further comparative measurements were conducted on selected soil samples. The results of these will be presented in section 3.4. The observation that the number of used PLS factors strongly influenced the differences between replicate measurements suggested that deviations in replicate analyses should – in addition to RMSE – be considered for selecting the optimum number of PLS factors (and probably also for comparing different calibration algorithms).

3.4. Reproducibility of VNIRS measurements

The reproducibility of VNIRS measurements and C_{org} estimates calculated thereof will be assessed in this section. Prior to presenting methods and results, the possible sources of uncertainties of VNIRS measurements and estimates will be outlined.

3.4.1. Sources of uncertainties

Errors and uncertainties of analytical results are caused by various sources. The very first is the question whether the collected soil samples represent correctly the units (area, horizon, time period, ...) they are representing. This is a fundamental problem of soil sampling, e.g. see Gruijter et al. (2006). Because the analytical methods considered for this work all used the same soil samples, it was not relevant for their comparison and will not be further discussed here.

For estimates of soil properties by VNIRS, possible error sources may be summarised in three groups:

- **Sub-sampling error** Are the analysed sub-samples representative for the entire soil samples?

- **Spectral errors** (random or systematic) are introduced by improper operation, by polluted white reference panels and petri dishes used for the measurements, and by errors of the spectrometer itself.

- **Model errors** are the inaccuracies of the model used to estimate the soil parameter of interest from the spectral data.

The first type of error, denoted as sub-sampling error, is introduced by variations within the collected soil material because usually only small portions ('sub-samples') of the entire soil samples are analysed. For smaller sub-samples as well as for more heterogeneous materials, larger sub-sampling errors are expected. Spectral errors may have various reasons, e.g. degraded panels for white reference, improper operation, or dysfunction of the spectrometer. While these problems are influencing directly the spectral data, their effects on the final results of the analyses – the estimates of the targeted soil property – are more complicated. Some spectral errors have no or only little effect on the final results, whereas other may strongly degrade the results. It was assumed that some calibration algorithms result in models more sensitive to errors in the spectral data than other. The model error is influenced by the accuracy of the reference analyses used for calibration. While systematic errors in the reference analyses are directly imported, random errors are leveled to some degree if enough samples are used (Naes, 2002).

3.4.2. Methods

To assess the reproducibility more exactly, ten soil samples were randomly selected and measured using the following setup: both devices (FS Pro and FS3) each equipped with a muglight contact probe were installed side by side. A sub-sample was taken from each sample and measured by both devices. After measuring all samples, a second sub-sample was taken from each sample and measured. A white reference measurement was taken after every three or four samples. After each white reference measurement, a sample consisting of fine sand was measured (the same sub-sample was used for all measurements). The sand sample was considered as standard measurement. The

3. VNIRS laboratory models

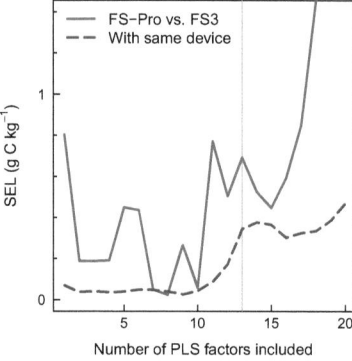 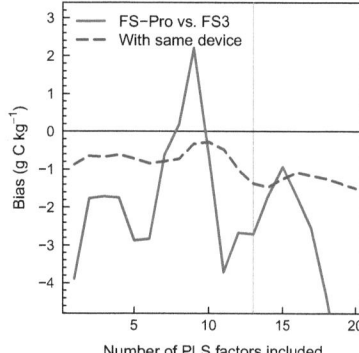

Figure 3.9 Standard error of laboratory (left) and bias of C_{org} estimates for duplicate analyses by the same VNIRS device as well as by different devices with respect to the number of used PLS factors.

spectra and the resulting C_{org} estimates were compared with respect to the two devices as well as with respect to the first and second measurement by the same device.

The standard error of laboratory (SEL) was calculated for VNIRS and compared to those of the used reference analyses. The SEL of the Swiss standard method (FAL, 1996) to estimate organic carbon was calculated using 14 duplicate analyses provided by Hauert (2007) and Ruth (2010). The SEL of total carbon determination by dry combustion (CN analyser) was estimated from 59 duplicate analyses of soil samples and 24 duplicate analyses of glutamic acid standards conducted within this project.

3.4.3. Results

Regarding the duplicate analyses conducted with the same VNIRS device, small SEL below $0.1 \, g \, C \, kg^{-1}$ were observed for models using 11 PLS factors or less (figure 3.9). Including more factors into the model resulted in strongly elevated SEL. For the models using 11 PLS factors or less, biases below $1 \, g \, C \, kg^{-1}$ were observed, while using more factors resulted in biases near $1.5 \, g \, C \, kg^{-1}$. The comparison of the scatter plots in figures 3.11 b and d indicated that the elevated SEL for models using a larger number of PLS factors was not only due to higher bias, but also due to increased variance.

Regarding the analyses of the same sub-samples by the two spectrometers, strong fluctuations in SEL and bias with respect to the number of used PLS factors were observed (figure 3.9). The smallest errors were observed for the models using 7 to 10 factors. For most models, the errors were bigger than those observed for the duplicate analyses conducted by the same device.

For the Swiss standard method according to FAL (1996), a SEL of $0.05 \, g \, C \, kg^{-1}$ was observed for C_{org} estimates. Compared to the mean of the used samples, this corresponded to a coefficient of variation (CV) of 0.04. For duplicate analyses of 59 soil samples by CN analyser, a SEL of $0.016 \, g \, N \, kg^{-1}$ and $0.14 \, g \, C \, kg^{-1}$ was observed for total N and C estimates, respectively. This

3.4. Reproducibility of VNIRS measurements

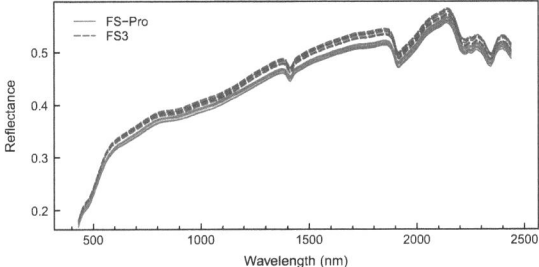

Figure 3.10 Reflectance spectra of fine sand as recorded by two different VNIRS devices (the same subsample of sand and the same white reference panel were used for all measurements).

corresponded to a CV of 0.05 and 0.04, respectively. Replicate analyses of 24 standard samples consisting of glutamic acid resulted in a SEL of $0.087\,g\,N\,kg^{-1}$ and $0.34\,g\,C\,kg^{-1}$, respectively, corresponding to CV below 0.01.

3.4.4. Discussion

It was not surprising that duplicate analyses conducted by different spectrometers resulted in higher SEL and particularly in higher bias compared to duplicate analyses by the same device. For the first, in addition to the common errors contributing to SEL, an additional error source was introduced by using two different spectrometers. The comparison of the spectra of the fine sand sample used as standard revealed that the two used spectrometers recorded slightly differing spectra (figure 3.10). The most remarkable difference was the generally higher albedo observed for the FS3 device. While slight deviations of the spectral curves were expected due to differences in the spectral resolutions of the two spectrometers, deviations with respect to the albedo were absolutely not expected because the same white reference was used for all measurements. The intention of the white reference measurements was to account and correct for differences introduced by differences in the geometry of the used contact probes, differences in the positions of the fiberglass optics, etc. Because of the observed differences for the sand sample, similar differences must be expected for the soil samples.

Apparently, the calibration of the spectral measurements by white reference measurements only was not sufficient. This procedure solely determined the amount of irradiation corresponding to 100 % reflectance and calibrated a linear relation between irradiation and reflectance based on this single observation. It was assumed that deviations from the expected linearity ocurred which were impossible to be recognised by this procedure. Presumably, the deviations depended on the spectrometer used. Possibly, an alternative procedure including several standards covering the full range of reflectance (e.g. 10, 25, 50, 75, 90, and 100 %) could account and correct for such deviations from linearity. Suitable reflectance targets are commercially available and it was wondered why – to my knowledge – such procedures were not used in soil spectroscopy. Although there was no proof for the assumed deviations from linearity, they seemed very probable as no other explanation was found for the observed deviations between the spectrometers.

It was investigated whether the differences in the spectra of the fine sand sample could be used to correct for the bias observed for the C_{org} estimates of the two devices. A first attempt to correct the spectra by applying the reflectance ratio FS Pro to FS3 for every wavelength failed because this procedure introduced a lot of features which degraded the C_{org} estimates (cf. figure A.16, page 117). Therefore, a constant ratio representing the ratio of the albedo of the sand spectra by

3. VNIRS laboratory models

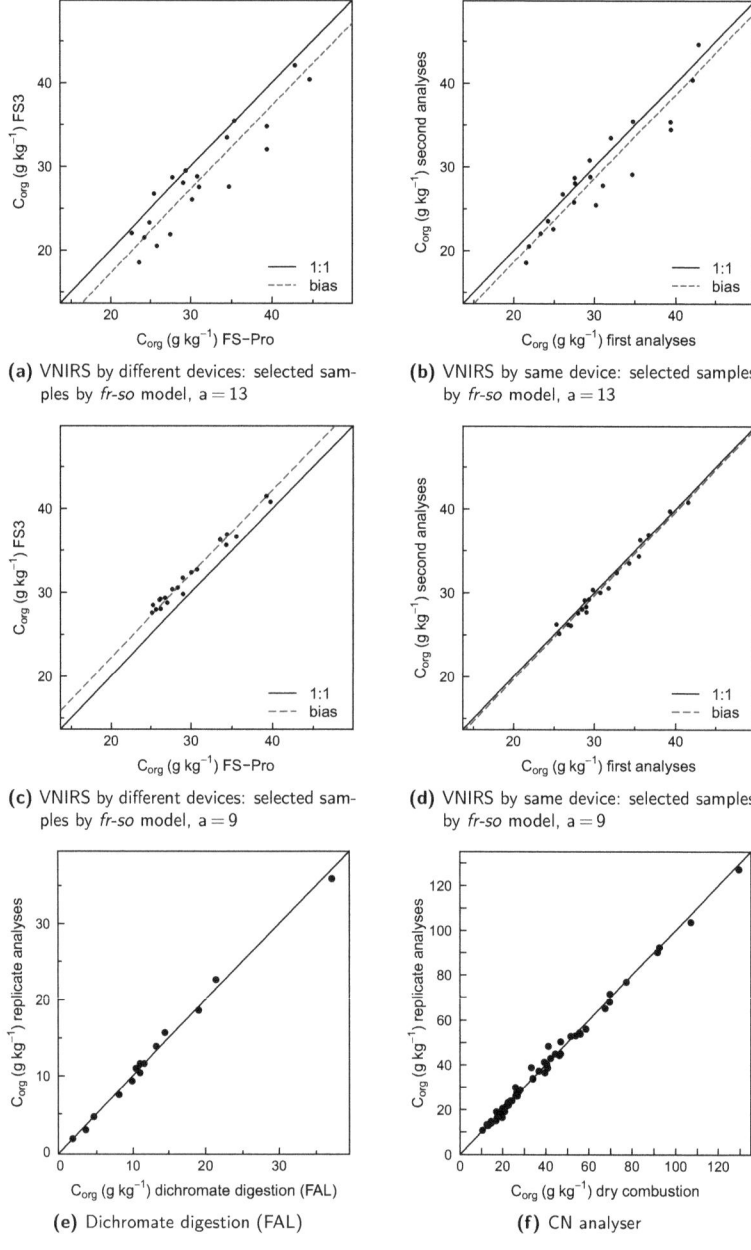

(a) VNIRS by different devices: selected samples by *fr-so* model, a = 13

(b) VNIRS by same device: selected samples by *fr-so* model, a = 13

(c) VNIRS by different devices: selected samples by *fr-so* model, a = 9

(d) VNIRS by same device: selected samples by *fr-so* model, a = 9

(e) Dichromate digestion (FAL)

(f) CN analyser

Figure 3.11 Reproducibility of C_{org} analyses by VNIRS, dichromate digestion according to FAL (1996), and dry combustion (CN-anaylser).

3.4. Reproducibility of VNIRS measurements

FS Pro and FS3 was used for the entire spectra. But also this approach did not result in considerable improvements of SEL or bias. Obviously, more complicated deviations relevant to the used PLSR models were present within the spectral data.

While the bias between duplicate analyses by different spectrometers was attributed to differences between the used devices, the origin of the observed bias between the first and second series of analyses by the same device remained unclear. Variations over time in the measurements might have been introduced by variations of the illumination or accidental displacements of the fiberoptical cable, but the white reference measurements should correct for these. In contrast, errors introduced by additional pollution of the white reference would not have been detected. The comparison of the spectra recorded for the sand sample used as standard during the first and second series of measurements revealed that the spectra of the second series exhibited mostly higher reflectance compared to those of the first series (figure A.15, page 117). These differences might have been random or systematic, e.g. due to pollution of the white reference panel. A t-test of the differences between the first and second series resulted in p-values between 0.01 and 0.05 (depending on the number of PLS factors included). Nevertheless, the observed differences were considered to be random. On one hand, the calculated p-values were not highly significant.[1] On the other hand, the two series of duplicate analyses were conducted within an interval of one hour and the panel used for white reference was sealed in a petri dish and handled with care, thus pollution was not probable.

Beside the observed biases with respect to time or the used spectrometer, also the variance within the measurements contributed to the observed SEL. The variance was assumed to be related mainly to sub-sampling errors, errors of the used petri dishes, and errors introduced by the white reference panel. Random errors of the spectrometer itself were considered of marginal importance as the noise observed for spectroscopic measurements was low compared to the variations found for replicate measurements of the used white reference panel (figure A.18, page 118). Also random errors introduced by the white reference panel seemed not to be dramatic because the deviations remained below 1 % reflectance. In contrast, systematic errors due to pollution of the spectralon can be dramatic as sugested by the evolution over time of the reflectance of the white reference panel used for this project (figure A.17). Back to the random errors: the contributions of the different sources of uncertainties to the variance observed for the VNIRS estimates could not be exactly identified, but in view of the low variance observed for the replicate analyses, this topic seemed of minor interest.

The presented and discussed results confirmed that with some models the reproducibility of VNIRS estimates of soil parameters was considerably higher than with other models. In other words: for some models, the random and systematic differences between the replicate analyses resulted only in slight differences, while for other models, they resulted in larger differences. The models compared in this study solely differed with respect to the number of PLS factors included. Thus, it was assumed that for completely different calibration algorithms, even more promoted differences may be observed. Therefore, it was recommended to consider the reproducibility during the calibration procedure. On one hand, the spectral measurements for a small portion (e.g. 5 or 10 %) of the calibration samples could be repeated and included as duplicates in the calibration dataset. This would result in models that are more stable with respect to random errors. On the other hand, SEL should be determined and used when comparing different calibration algorithms, but also when fine-tuning the models, e.g. for selecting the optimum number of PLS

---

[1] In contrast, the differences between the two spectrometers were highly significant ($p \ll 0.001$).

factors. It was suggested that using SEL in addition to RMSE would further enhance the stability of the derived models.

Comparing visually the reproducibility of the reference methods and VNIRS estimates revealed that good VNIRS models were able to compete with the reference methods, while bad models exhibited clearly worse reproducibility (figure A.14). When considering the coefficient of variation (CV) of SEL, which describes the relative error with respect to the mean of the data, comparable CV of 0.04 and 0.05 were reported for C_{org} estimates of the considered reference methods. For VNIRS estimates, clearly lower CV were reported: the good models achieved CV between 0.001 and 0.002 when using the same spectrometer for duplicate analyses. When using worse models or duplicates by different spectrometers, CV increased to values between 0.01 and 0.02. Obviously, VNIRS analyses were clearly better reproducible. Most of the differences in reproducibility were attributed to the fact that for VNIRS measurements much bigger sub-samples (15-20 g) were used compared to the reference methods (0.1-0.2 g). Variations within the soil samples were leveled by the bigger samples. The assumption that the variation within the soil samples was influent for small sub-samples was supported by the observation that the CV of glutamic acid (being a very homogeneous material) was much lower compared to the CV of soil samples analysed by CN analyser.

3.5. Summary and Conclusions

Within this chapter, the ability of the PLSR algorithm to estimate C_{org} and N_{tot} contents from VNIR spectra collected from dried and sieved samples was assessed. Different sets of soil samples covering local to regional to national areas were considered. The models were validated by six-fold cross-validation. Additionally, models derived from a regional dataset were used to estimate C_{org} for a set of validation samples originating from a neighboured region. The reproducibility measured as standard error of laboratory (SEL) of duplicate analyses as well as the additional errors introduced by using two different spectrometers were assessed.

For C_{org} as well as N_{tot}, very good results were achieved for datasets of local to regional extensions. The observed cross-validation RMSE ranged from 1.9 to $2.7\,g\,C\,kg^{-1}$ and from 0.19 to $0.24\,g\,N\,kg^{-1}$, respectively, and were comparable to RMSE reported in published studies. Considerably higher RMSE resulted for the national datasets ($9.8\,g\,C\,kg^{-1}$ and $0.63\,g\,N\,kg^{-1}$, respectively). The bad performances compared to published studies were explained by the rather small number of soil samples compared to the variability of soils covered by the national datasets. The results for the validation of the model to estimate C_{org} showed good correlations with the reference analyses, but were biased.

The VNIRS analyses exhibited low SEL ($< 0.1\,g\,C\,kg^{-1}$). VNIRS achieved lower CV_{SEL} than the used reference methods (0.001 to 0.002 vs. 0.04 to 0.05). The RMSE as well as SEL strongly depended on the number of PLS factors included. Therefore, it was suggested that both should be considered for comparing different calibration algorithms and for fine-tuning the used models.

3.4. Reproducibility of VNIRS measurements

Based on the presented results, it was concluded:

- Given a suitable calibration dataset is available, C_{org} and N_{tot} can be reliable and accurately estimated by VNIRS.

- In order to achieve accurate predictions, specific calibrations for homogeneous sets of soil samples (selected region, selected horizon, ...) are desirable.

- Calibrations derived from heterogeneous sets of soil samples can be adapted for homogeneous sets of samples by determining the bias of the estimates.

- The stability of the resulting models should be considered during selection and fine-tuning of models. SEL was suggested as suitable proxy.

- Score and orthogonal distances should be further assessed as potential tools to judge the applicability of VNIRS calibrations on unknown soil samples.

- The observed differences between reflectance spectra collected by different spectrometers are unsatisfactorily as they introduce significant errors for the resulting estimates of soil properties.

- To guarantee a stable performance on the long-term and between different spectrometers, the currently used procedure for white reference should be revisited and additional standardisation mechanisms should be considered.

4. Comparing different calibration algorithms using wavelets[1]

During the last decade, Visible and Near Infrared Reflectance Spectroscopy (VNIRS) has experienced a boom and is considered now as a promising tool for soil analysis, allowing for rapid and inexpensive measurements of soil properties like the contents of organic carbon, nitrogen, iron oxides, clay and others (Stenberg et al., 2010). The rising popularity of VNIRS has been accompanied by a large number of publications (e.g. see table 1 in Viscarra Rossel et al., 2006a) as well as extensive discussions about calibration algorithms that should be used. Because soil absorption features in the visible and near infrared are mostly unspecific, very broad and overlapping, statistical models are needed to connect the spectral data to the soil property analysed (Workman, 2008). Partial least squares regression (PLSR) is the most popular technique and used for the majority of the VNIRS calibrations in soil science, but does not always result in accurate estimations and hardly accounts for non-linearities (Naes, 2002).

Wavelet transforms decompose each spectrum into a sum of wavelets that reproduce a unique pattern at different scales and positions (Lark & Webster, 1999). A coefficient is calculated for every scale and position of the wavelet and reflects its contribution to the sum (figure 4.1). The shape of the wavelet remains constant, but its width is halved going from one scale to the next finer, and within every scale its position is shifted along the x-axis. Therefore, coarse scales consist of few wavelet coefficients covering broad areas of the spectrum while fine scales consist of many wavelet coefficients each representing a narrow area of the spectrum (see section 2.5.6 for more details). Because a relatively small number of wavelet coefficients contains most of the original information, spectral data are compressed and denoised by selecting a subset of wavelet coefficients (Alsberg et al., 1997b). Viscarra Rossel & Lark (2009) ordered the wavelet coefficients according to their variance (from highest to lowest) and selected coefficients with high variances for their models. The authors argued that the variance was a measure of the information present in each wavelet coefficient.

The objective of the work presented within this chapter was to estimate total nitrogen (N_{tot}) and organic matter (OM) contents of soil samples from their VNIR spectrum using wavelet transforms based on the algorithm of Viscarra Rossel & Lark (2009) and adapted from it: instead of variance, we used covariance, Pearson's and Spearman's correlation coefficients, and the median absolute deviation (MAD) to order and select the wavelet coefficients. The selected subsets of wavelet coefficients were then used for regression with quadratic polynomials. The performance of wavelet transforms with these algorithms was addressed for OM and N_{tot} through cross-validation and validation; it was also compared to that of PLSR.

---

[1] The content of this chapter was elaborated in cooperation with Bernard Barthès, IRD Montpellier. I would like to thank Bernard whose inputs and recommendations improved this chapter substantially.

4. Comparing different calibration algorithms using wavelets

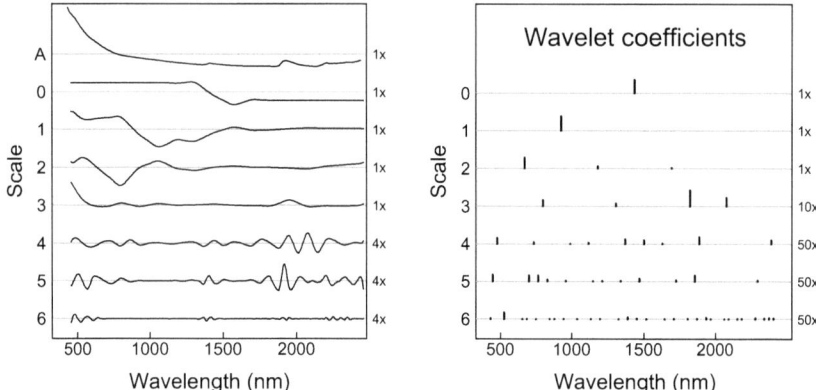

Figure 4.1 Example of wavelet decomposition for one randomly selected soil sample: absorbance spectrum of the soil sample (A), its decomposition into wavelets of single scales from coarse to fine scales (0 to 6 on left side), and corresponding wavelet coefficients (right side; line length is proportional to value of corresponding wavelet coefficient). The two finest scales, 7 and 8, are not shown. Right hand axes indicate the magnification factors used for plotting of the different scales.

4.1. Materials and methods

4.1.1. Soil samples and reference analyses

Two different sets of archived soil samples were used to derive VNIRS estimation models for organic matter (OM) and total nitrogen (N_{tot}) contents (table 4.1). The 567 soil samples used for the N_{tot} models originated from 115 forest sites covering whole Switzerland and were provided by the Swiss Federal Institute for Forest, Snow and Landscape Research (WSL), Birmensdorf, Switzerland. The soil samples had been collected by horizons over the whole profile. The 704 samples used for the OM models originated from 469 sites in north-western Switzerland (Cantons of Fribourg and Solothurn) and were provided by the Soil Protection Unit of the Canton of Solothurn and the Soil Monitoring Unit of the Canton of Fribourg (FRIBO). Approximately 80 % of these had been taken as mixed samples from the top soil layer (0-20 cm), and the remaining ones had been collected by horizons within the 0 to 30 cm depth layer. The samples of both sets were stored dried and sieved (<2 mm).

The models for OM were validated using an independent set of 54 soil samples collected by the University of Bern at two villages of north-western Switzerland (Reisiswil and Gondiswil). The two villages were close to, but not included in the calibration area, and exhibited soils comparable to those of the calibration area. N_{tot} models were not validated.

The analyses for N_{tot} and OM were provided by the owners of the soil archives. The N_{tot} analyses were done by dry-combustion using a CN analyser (CE Instruments NA 2500, Wigan, UK). The OM analyses were conducted according to the Swiss Standard Method (FAL, 1996), which involves dichromate oxidation.

4.1. Materials and methods

Table 4.1 Studied sets of soil samples for wavelet models. n_{total}: number of samples in the original set; n_{out}: number of spectral outliers; n_{used}: number of samples used for calibration and validation; sd: standard deviation; statistics refer to the used soil samples.

Dataset	n_{total}	n_{out}	n_{used}	Min.	Median	Mean	Max.	sd
					— (g kg^{-1}) —			
N_{tot} – calibration	653	86	567	0.0	0.8	1.5	9.7	1.8
OM – calibration	789	85	704	6	27	31	95	15
OM – validation	54	5	49	15	28	31	63	11

4.1.2. Data collection

The spectral data were collected using an ASD FieldSpec 3 High Resolution (ASD Inc., Boulder, CO, USA) with a muglight contact probe and a Spectralon panel for white reference. The measuring procedure described in section 2.3 was used: the soil samples (10 to 15 g) were filled into Petri dishes and the reflectance was measured through their bottom side. Reflectance was collected from 350 to 2500 nm, averaging 20 co-added scans per measurement, and measuring every sample twice after rotating the sample by 90° and using the mean of the two measurements. Reflectance values at intervals of four nm were used for the data analysis. The data range was reduced to 412-4256 nm to remove areas with a low signal-to-noise ratio and to make the data suitable for wavelet decomposition (length of data must be power of two). Furthermore, the steps occurring at the sensor changes were removed by shifting the different segments to form a continuous line. Reflectance spectra (R) were converted into absorbance spectra (A) by $A = -\log R$.

The datasets were tested for spectral outliers by principal components analysis. Based on the Mahalanobis distance MD using the first seven principal components, samples with $MD^2 > \chi^2_{0.95,7} = 14.1$ (the quantile of the chi-squared distribution for seven degrees of freedom) were considered outliers and removed for all models (De Maesschalck et al., 2000). Table 4.1 presents the datasets after outlier removal. The number of samples removed as spectral outliers seemed to be exorbitant, but the original sets of samples included a large number of extreme soil samples (e.g. samples from very deep soil layers, from mountainous sites with extreme conditions, etc.). Without removing these extreme soil samples, it would have been very difficult interpreting and comparing the results.

4.1.3. Algorithms and models

Throughout this chapter, the term *algorithm* refers to the mathematical procedure used for calibration (to link the spectral data with the soil parameter considered) while the term *model* refers to the mathematical equation produced by the calibration algorithm.

The calibration algorithms used in this work were adapted from an algorithm presented by Viscarra Rossel & Lark (2009): they decomposed the spectral data using wavelet transforms and ordered the wavelet coefficients according to their variance, from the highest to the lowest. They started with one coefficient and added then one by one to their models until the difference between measured and estimated values in cross-validation, expressed as root mean squared error, reached

4. Comparing different calibration algorithms using wavelets

its minimum. They achieved the best results using multiple linear regressions with quadratic polynomials applied to spectral data reconstructed from the selected wavelet coefficients.

In the present work, wavelet coefficients were calculated from absorbance data using Daubechies' so-called extremal phase wavelet number eight (Nason, 2008). To avoid problems near the ends of the spectra, the modified wavelet functions proposed by Cohen et al. (1993) were used (for details on the wavelet decomposition, see Lark & Webster, 1999, Nason, 2008, or section 2.5.6 of this work).

For each algorithm, we started with two wavelet coefficients and added coefficients one by one up to 50 coefficients. We alternatively used variance, covariance, Pearson's and Spearman's correlation coefficients, and the median absolute deviation MAD) to select the wavelet coefficients. Covariance and correlation coefficients were calculated between each wavelet coefficient and the target variable. The MAD is a robust statistic for the variance and for a univariate dataset $x = x_1, x_2, \ldots, x_n$ it is calculated as:

$$\text{MAD} = \text{median}\left[x_{i=1\ldots n} - \text{median}(x)\right] \tag{4.1}$$

The selected wavelet coefficients were used in a linear regression model including linear and quadratic terms:

$$y = \alpha + \beta_1 x_1 + \gamma_1 x_1^2 + \beta_2 x_2 + \gamma_2 x_2^2 \ldots + \beta_n x_n + \gamma_n x_n^2 \tag{4.2}$$

where y represents the target variable, n the number of used wavelet coefficients, x_1 to x_n the selected wavelet coefficients, and α, β_i and γ_i the model parameters. According to Viscarra Rossel & Lark (2009), it is possible to use wavelet coefficients directly in linear regressions because the wavelet coefficients are decorrelated.

The performances of the different wavelet models were compared one to another; they were also compared with the performance of partial least squares regression models (PLSR; Bjørsvik & Martens, 2008). As for the wavelet models, we used two to 50 PLS factors for the models.

All calculations were computed using the statistical software R (version 2.10.0) and the package wavethresh (Nason, 2008) for the wavelet decomposition.

4.1.4. Model validation

The considered algorithms were tested by six-fold cross-validation. The groups for cross-validation were selected ensuring that all soil samples from a given sampling site belonged to the same group, in order to improve model robustness. For the calibration, the accuracy of the estimations was assessed using root mean squared error (equation 2.22, page 23) of cross-validation. Because we found that the performance of the models varied considerably when using different cross-validation segmentations, we decided to repeat the cross-validation ten times and to report the average RMSE. For all models, the same cross-validation segmentations were used. For the validation, the accuracy of the estimations was assessed using the RMSE of the validation samples.

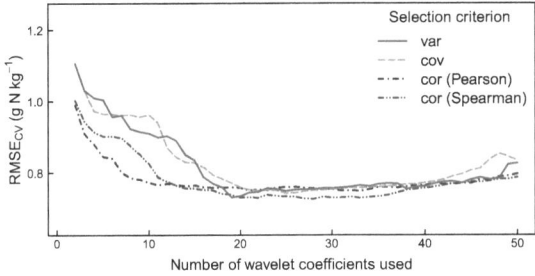

Figure 4.2 Comparison of calibration RMSE for N_{tot} estimates by algorithms using variance (var), covariance (cov), and correlation coefficients (cor) as selection criterion for wavelet coefficients (average of 10 replicates).

4.2. Results

4.2.1. Wavelet model performance

All algorithms succeeded to derive models to estimate N_{tot} and OM from VNIR spectra. The wavelet models estimated N_{tot} with minimal RMSE of cross-validation between 0.73 and 0.75 g kg^{-1} (figures 4.2 and 4.3). The best RMSE (0.73 g kg^{-1}) was achieved when using either 19 wavelet coefficients selected by variance or 28 coefficients selected by Spearman's correlation coefficient. For all models, RMSE of cross-validation first decreased when adding wavelet coefficients, then remained stable over a wide range, and started to increase again when using more than 40 coefficients. The algorithm using Pearson's correlation coefficient reached its stable range of RMSE when using 10 wavelet coefficients or more, while the other algorithms needed about 20 coefficients. The performance of the model using MAD was very similar to that of the model using variance, but the former needed 23 wavelet coefficients to reach its optimum performance and the latter 19.

All wavelet models estimated OM with a minimum RMSE of cross-validation between 5.4 and 5.7 g kg^{-1} (figures 4.4 and 4.5). The best RMSE (5.4 g kg^{-1}) was achieved when using 38 wavelet coefficients selected by covariance. The models using covariance and correlation coefficients as selection criterion showed very similar performances with a decrease in RMSE from about 8 to 5.6 g kg^{-1} when adding the first 20 to 25 wavelet coefficients. Beyond this point, the models performance changed only slightly. The models using variance and MAD as selection criterion had higher RMSE when using 10 to 30 wavelet coefficients. When using 20 to 30 coefficients, they achieved a RMSE of cross-validation around 6.4 g kg^{-1}. To achieve performances comparable to the other models, 43 or more coefficients were needed, resulting in a RMSE of 5.7 g kg^{-1}.

The validation RMSE fluctuated strongly within the considered range of wavelet coefficients (figures 4.6 and 4.7). The models using variance showed the most stable results with RMSE between 6.2 and 8.2 g OM kg^{-1} over the whole range and a stable region from 20 to 27 wavelet coefficients with a RMSE of 7.0 g kg^{-1}. The best RMSE (6.2 g kg^{-1}) was achieved when using only two coefficients. The models using covariance showed results comparable to those using variance below 30 coefficients, but RMSE increased considerably beyond 30 coefficients. The models using correlation coefficients showed good results when using 13 to 16 wavelet coefficients, with RMSE between 7.0 and 7.5 g kg^{-1}, but clearly higher RMSE outside this range. The results of the models using MAD showed results very similar to those using variance.

Figure 4.8 illustrates the scale and position of the wavelet coefficients selected by the different criteria used with our algorithms. Using variance and MAD promoted coefficients of the coarse

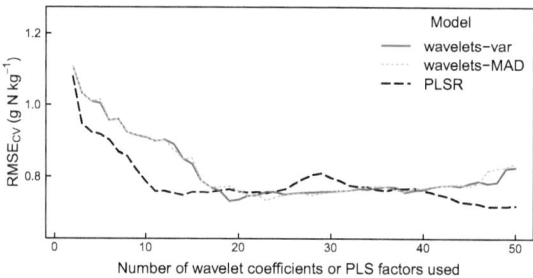

Figure 4.3 Comparison of calibration RMSE for N_{tot} estimates by algorithms using variance (var) and median absolute deviation (MAD) as selection criterion for wavelet coefficients, and with PLSR (average of 10 replicates).

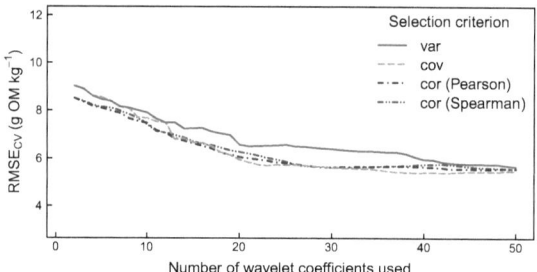

Figure 4.4 Comparison of calibration RMSE for OM estimates by algorithms using variance (var), covariance (cov), and correlation coefficients (cor) as selection criterion for wavelet coefficients (average of 10 replicates).

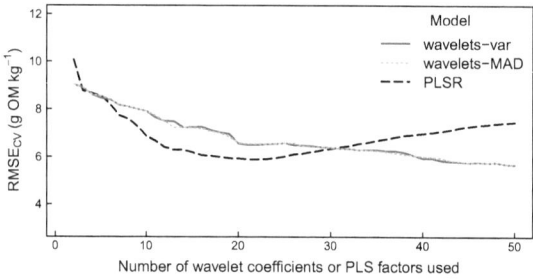

Figure 4.5 Comparison of calibration RMSE for OM estimates by algorithms using variance (var) and median absolute deviation (MAD) as selection criterion for wavelet coefficients, and with PLSR (average of 10 replicates).

4.3. Discussion

scales (scales zero to three) for both N_{tot} and OM. Covariance also promoted coefficients of the coarse scales, but with a slight shift towards the middle scales (scales three to five). Finally, the algorithms using correlation coefficients mainly promoted coefficients of the fine scales (scales six to eight) for N_{tot} and coefficients of the middle scales for OM.

4.2.2. Comparison with PLSR

The PLSR models for N_{tot} showed good performance with an RMSE of cross-validation near 7.5 g kg^{-1} when using between 11 and 25 PLS factors (figure 4.3). When using more factors, RMSE of cross-validation increased to reach a maximum near 30 factors then decreased again. When using 50 PLS factors, RMSE of cross-validation fell slightly below 7.5 g kg^{-1}.

The PLSR models for OM achieved their optimum performance when using 22 factors (figure 4.5). The RMSE of cross-validation reached 5.9 g kg^{-1} and was slightly higher than those of the wavelet models. When using 15 to 20 PLS factors, RMSE was still good (between 6.0 and 6.1 g kg^{-1}). When using more than 25 factors, RMSE increased considerably.

As regards the validation of the OM models (figure 4.7), PLSR achieved a RMSE of 6.1 g kg^{-1} when using 12 PLS factors. This was better than for models based on wavelet coefficients. When using up to 20 factors, RMSE was fluctuating between 6.1 and 7.5 g kg^{-1}, and increased rapidly beyond 20 factors.

4.3. Discussion

The N_{tot} estimation models presented in this work all achieved best cross-validation RMSE near 0.75 g N kg^{-1}. Using NIRS, Chang et al. (2001) were able to estimate N_{tot} of soil samples from the top 30 cm with a RMSE of 0.62 g kg^{-1} by selecting spectrally similar calibration samples and applying principle component regressions. Using NIRS and PLSR, Brunet et al. (2007) were able to estimate N_{tot} with a RMSE of cross-validation between 0.20 and 0.26 g kg^{-1} for < 2-mm sieved samples, depending on the mathematical pretreatment. They even reached much lower RMSE by using subsets based on texture or geographic origin and by sample grinding (< 0.2 mm). Brunet et al. (2007) used soil samples from eight tropical sites, which had much lower N_{tot} contents (mean 0.6 g N kg^{-1}; maximum 2.3 g N kg^{-1}) than those of the present work (mean 1.5 g N kg^{-1}; maximum 9.7 g N kg^{-1}). It is hypothesized that higher N_{tot} contents caused higher RMSE here. The soil samples used by Chang et al. (2001) had N_{tot} contents comparable to those used in the present work, but we assumed that our collection of soil samples was more diverse because we used samples from the whole soil profile and not only the top layer. This could explain the slightly higher RMSE we achieved.

Stenberg et al. (2010) compared 21 published studies presenting OM estimations by VNIRS. They reported large variations in RMSE of cross-validation between different studies and showed that RMSE of cross-validation strongly correlated to the standard deviation of the sample set. Based on their compilation (Stenberg et al., 2010, cf. figure 3.2, page 31) and on the standard deviation of our sample set (15 g OM kg^{-1}), a RMSE near 5 g kg^{-1} could be expected. The achieved RMSE between 5.4 and 5.7 g OM kg^{-1} matched the expectations very well, even though very strict restrictions had been applied for cross-validation (all samples from the same site belonged to the same segment), which tended to increase the RMSE.

For N_{tot} estimations, the algorithms using correlation coefficients as selection criterion showed

4. Comparing different calibration algorithms using wavelets

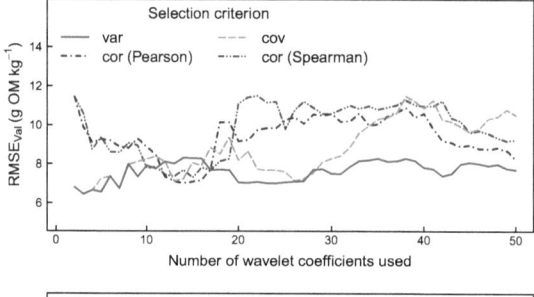

Figure 4.6 Comparison of validation RMSE for OM estimates by algorithms using variance (var), covariance (cov), and correlation coefficients (cor) as selection criterion for wavelet coefficients.

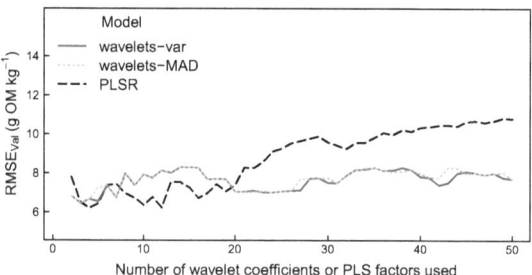

Figure 4.7 Comparison of validation RMSE for OM estimates by algorithms using variance (var) and median absolute deviation (MAD) as selection criterion for wavelet coefficients, and with PLSR.

a faster decrease in cross-validation RMSE for small numbers of wavelet coefficients than the other wavelet algorithms. However, when using 20 coefficients or more, all algorithms performed similarly. Obviously, using correlation coefficients promoted the selection of wavelet coefficients closely related to the soil property to be estimated. We assumed that using variance, covariance and MAD promoted wavelet coefficients that represented dominant variations in soil characteristics, either correlating or not with the property of interest.

As regarded cross-validation for OM, the decrease in RMSE with increasing number of wavelet coefficients was comparable for the models using correlation coefficients and covariance. The decrease was similar for the models using variance and MAD but shifted to slightly higher RMSE.

There were two approaches to explain why the algorithms using correlation coefficients resulted in more parsimonious models than the other wavelet algorithms for N_{tot}, but not for OM: differences between the estimated soil properties, and differences between the sets of soil samples. Organic matter has a very strong influence on the soil spectra; it has an impact over the whole visible and near infrared range and is in fact one of the most important (if not the most important) characteristic governing soil reflectance. By contrast, the influence of the chemical compounds related to N_{tot} is minor and – if visible – restricted to smaller areas. Often, it has even been claimed that VNIRS estimations of N_{tot} were not at all using spectral features caused by N-bonds but correlations with other soil properties (Stenberg et al., 2010). As a matter of fact, it seemed that dominant variations in soil properties (particularly promoted through selection according to variance and MAD) often coincided with wavelet coefficients important for OM estimation but not for N_{tot} estimation. The second approach focused on differences between the sets of soil samples: the set used for OM was much more homogeneous (regional scale, samples from the top 30 cm) than that used for N_{tot} (national scale, samples from the whole soil profiles). Thus, the OM set exhibited less spectral differences, for instance due to mineralogy. Both effects might have contributed to the different

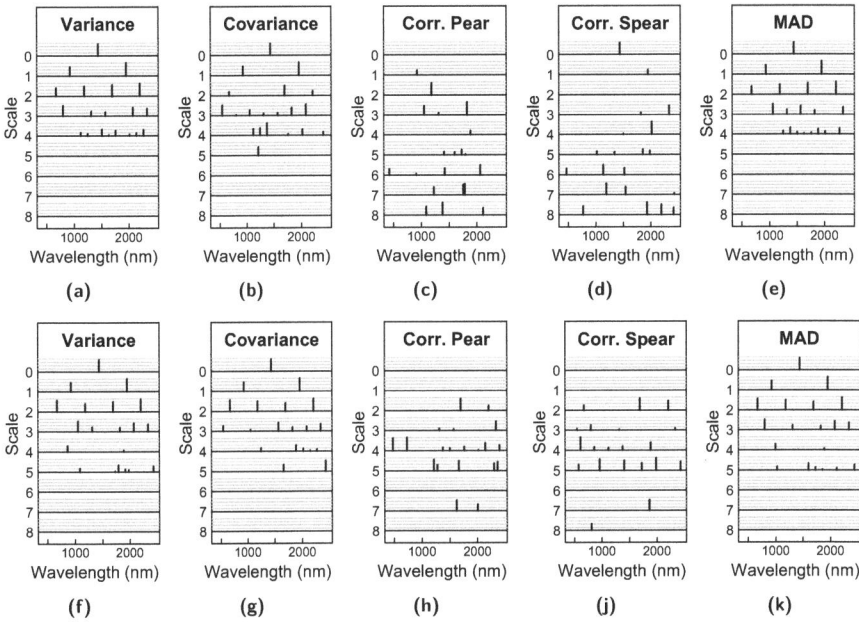

Figure 4.8 Scale and position of the first 20 wavelet coefficients selected by the algorithms using variance, covariance, correlation coefficients, and MAD for (a-e) N_{tot} and (f-k) OM estimations. The wavelet coefficients corresponding to the longest bar were selected first, those corresponding to the smallest bar in the twentieth place.

behaviour of OM and N_{tot} models. By contrast, the reasons why the models using variance and MAD provided worse OM estimations than the others remained unclear. Possibly, some of the wavelet coefficients selected according to variance and MAD represented noise or unimportant information.

Using selection criteria such as Spearman's correlation coefficient and MAD – which are more robust than Pearson's correlation coefficient and variance, respectively – did not improve or degrade the models. This might be due to the fact that spectral outliers had been removed a priori using relatively strict criteria. Studying datasets containing outliers would probably enhance the performance of models based on robust selection criteria.

Actually, the different selection criteria used with the algorithms included very different subsets of wavelet coefficients (figure 4.8). While using variance, covariance and MAD promoted the inclusion of coarse scale coefficients, using correlation coefficients promoted middle and fine scale coefficients. This could support the above-mentioned assumption that variance, covariance and MAD mainly extracted information related to the dominant variations within the set of soil samples, while correlation coefficients extracted information more closely related to the soil property of interest. Using coefficients of correlation for selection, the greater importance of wavelet coefficients from the fine scale for the N_{tot} than for the OM models evidenced that chemical compounds related to N_{tot} had narrower and more specific influence on spectra than those related to OM.

The OM validation results exhibited large variations in RMSE depending on the number of wavelet coefficients used. Because the validation soil samples all originated from a small area, they represented a much more homogeneous set than the calibration samples. As a consequence, they were either all accurately predicted when the number of wavelet coefficient was appropriate, or all poorly predicted otherwise. It seemed that the algorithms using correlation coefficients were prone to extract spectral information that was not present or not important for the validation samples, as they often resulted in poor OM validation. Nevertheless, they yielded good RMSE when using 13 to 16 wavelet coefficients. The problem is that this range of coefficients would probably not be chosen, because the cross-validation led to use about 20 coefficients. Based on the validation data, using variance and MAD seemed to be the most relevant solution because its validation RMSE was more stable than those of the other selection criteria; and because it showed good performance at the optimal range of wavelet coefficients indicated by the cross-validation. It might however be assumed that using correlation coefficients would be more successful for validation sets spectrally closer to, thus better represented by, the calibration set.

For both target variables, PLSR yielded results comparable to those of wavelet algorithms. This led to the conclusion that wavelet decomposition was as effective as PLSR to compress and denoise spectral data. However, it might be hypothesized that removing spectral outliers by principal component analysis favoured PLSR over wavelet algorithms, because principal component analysis and PLSR are comparable approaches. Outlier removal based on wavelet coefficients would possibly result in better results for wavelet algorithms than for PLSR. Moreover, considering PLSR validation results, it seemed more relevant to choose the optimal number of PLS factors according to slowdown decrease than to minimum cross-validation RMSE. Here, this would be near 12 factors instead of the 20 factors indicated by minimum RMSE. The additional PLS factors might improve cross-validation performance by extending the model, but they also enhanced overfitting effects leading to poorer validation performance.

4.4. Summary and Conclusions

Amongst others, models based on wavelet transforms have been used to analyse soil spectral data. Wavelet transforms decompose each spectrum into a sum of wavelets that reproduce a unique pattern at different positions and scales, and are represented by wavelet coefficients. In the work presented within this chapter, five adaptations of an algorithm presented by Viscarra Rossel & Lark (2009) were compared to estimate N_{tot} and OM contents from VNIR spectra; the performance of partial least square regression (PLSR) was also addressed. The adapted algorithms used variance, covariance, Pearson's and Spearman's correlation coefficients, and median absolute deviation (MAD) to select a subset of wavelet coefficients that was used for regression with quadratic polynomials. Two sets of archived soil samples from Switzerland were studied. For both sets, six-fold cross-validation was conducted. In addition, a set of independent validation samples was used for OM.

In cross-validation, the algorithms based on wavelet coefficients achieved RMSE consistent with those of published works (0.73 g kg^{-1} and 5.5 g kg^{-1} for N_{tot} and OM, respectively). For N_{tot}, using correlation coefficients to select wavelet coefficients produced more parsimonious models than using variance and covariance. For OM, differences between algorithms were negligible. Algorithms using correlation coefficients promoted wavelet coefficients from middle and fine scales, while the others promoted coefficients from coarse scales. The models using variance seemed more stable and reliable, especially for validation. Similarly accurate estimations were achieved using PLSR.

Based on the presented results, it was concluded:

- The algorithm presented by Viscarra Rossel & Lark (2009) was suitable to estimate soil parameters from spectral data, especially for large datasets.

- Although for some situations other algorithms were more successful, we recommend variance (or its robust counterpart MAD) as selection criterion for wavelet coefficients.

- Further research is needed to assess the circumstances under which using correlation coefficients would be favourable and how the selection of wavelet coefficients could be improved, e.g. by expert knowledge.

5. Soil moisture effects on VNIR reflectance

It is a common observation familiar to most soil scientists, farmers, and gardeners that soil appears darker upon wetting. This implies that the reflection of light is decreased and the absorption is increased by the additional water. It seems obvious that this effect is not restricted to visible radiation, and reflection in the infrared is altered, too. In order to use VNIR spectroscopy directly in the field or to measure field moist soil samples, it is an indispensable prerequisite understanding the influence of soil moisture on reflectance. For example, organic matter content strongly influences soil darkness, too. Hence, it must be expected that soil moisture influences VNIRS estimates of organic matter contents. Therefore, there is an interest in finding the relations between soil moisture and reflectance which, possibly, would allow to correct for the interferences by soil moisture. Of course, the same relations could also be used to measure soil moisture. There are other fast and easy-to-use techniques to measure soil moisture – e.g. TDR –, but VNIRS would be a good choice to capture the moisture of the soil surface. Within this work, a diverse set of 43 soil samples was measured at different moisture contents from saturated to oven dry. Parameters including reflectance for defined wavelengths, reflectance ratios, and integrals of parts of the spectra were tested for their adequacy for soil moisture estimation. Special attention was paid to the question whether the relation between the parameters and soil moisture could be generalised for different soils.

5.1. Water absorption features

All absorption bands attributed to the three fundamental vibrations of the H_2O molecule in liquid water occur in the MIR near 3100 nm (symmetric O-H stretch), 6080 nm (H-O-H bend), and 2900 nm (asymmetric O-H stretch). The O-H stretch is responsible for the reflectance decrease visible in the NIR beyond 2300 nm (Clark, 1999). Combinations of these are visible in the NIR (figure 5.1), whereas overtones also occur in the same range, but are hardly seen because they

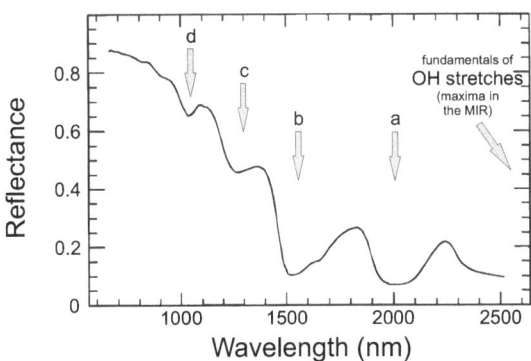

Figure 5.1 NIR spectrum of solid water (modified from Clark, 1999) and the most important absorption features: combinations of (a) H_2O bend and asymmetric OH stretch, (b) symmetric and asymmetric OH stretch, (c) H_2O bend and both OH stretches, and (d) asymmetric OH stretch and the first overtone of the symmetric OH stretch.

5. Soil moisture effects on VNIR reflectance

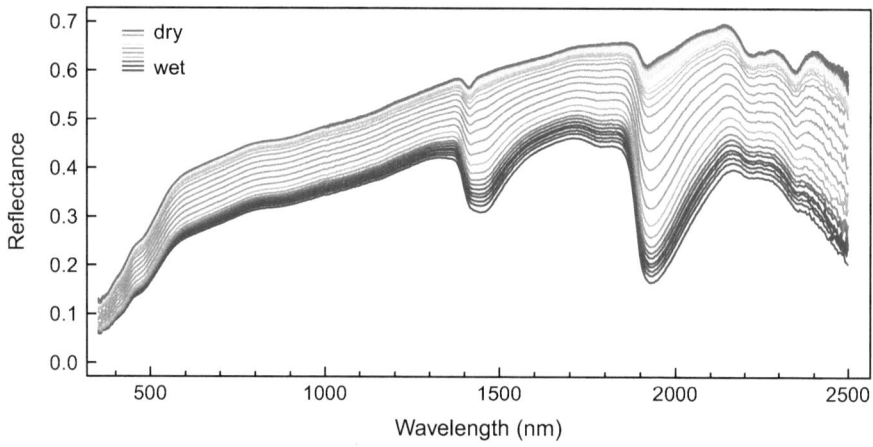

Figure 5.2 Changes in reflectance of sand during drying out.

are much weaker (Workman & Weyer, 2007). At room temperature, the two most prominent absorptions appear near 1450 and 1940 nm (around 1500 and 2000 nm for ice, respectively), while weaker absorption bands near 740, 840, 1200, 1780 nm are observed. Generally, absorption maxima are shifted to lower wavelengths with increasing temperature due to changes in hydrogen bonding (Workman & Weyer, 2007).

Similarly, spectra of dry soil samples exhibit two prominent absorption bands near 1450 and 1940 nm attributed to water and hydroxyl groups incorporated in the mineralogical structure. The first was identified as combination of symmetric and asymmetric O-H stretches. Thus, the absorption band around 1450 nm may be caused by H_2O and/or O-H groups. In contrast, the band around 1940 nm was identified as combination of H-O-H bending and asymmetric O-H stretch. Therefore, the presence of this absorption band doubtlessly indicates the presence of H_2O molecules in the soil. Broad water absorption bands indicate that H_2O molecules are present in different locations/configurations and that they are little ordered (Clark, 1999).

In addition to the so-called structural water, moist soil samples exhibit water films in the pore space and on the aggregate surfaces. Thickness and distribution of these layers depend on the water content as well as physical and chemical soil properties. Figure 5.2 illustrates the effect of increasing water contents on soil reflectance: the reflectance is decreased over the whole wavelength range – or in other words: the albedo is decreased –, the strong absorption bands near 1450 and 1940 nm are getting deeper and broader while its absorption maxima are slightly shifted to higher wavelengths, and a weaker absorption band near 1780 nm is appearing. Moreover, the shape of the spectra beyond 1400 nm is getting more and more tilted to the right hand side, and the absorption features above 2000 nm are progressively degraded.

The expansion of the two strong water bands for increasing water contents are clearly attributed to the mechanisms explained above. The broadening also indicates that the H_2O molecules are becoming less ordered with increasing moisture contents. The reduction in albedo is usually explained by changes in the index of refraction due to thin water films and dissolved soil constituents therein as well as changes in the physical properties of the soil particles due to water (Baumgardner et al., 1985). The last two effects – tilting to the right hand side, and degrading absorption features – are caused by the absorption of the asymmetric O-H stretch (fundamental vibration) centred around 2800 nm. (Clark, 1999).

Bowers & Hanks (1965) reported good fits between reflectance values and soil moisture with the restriction that the moisture-reflectance curve was unique for each soil. They recommended the 1900 nm band for soil moisture prediction because its greater sensitivity to water, although other wavelengths could be used. These results were confirmed by Skidmore et al. (1975) who found log-linear relationships for 1950 nm reflectance vs. moisture that differed between soils. The effect that the overall shape of the reflectance curves are inclining to the right hand side was used by Whiting et al. (2004) to estimate gravimetric soil water contents of samples from two Mediterranean regions. They fitted an inverted Gaussian function to the normalised soil spectra and used the area under this curve to estimate water content. When using only samples exhibiting water contents below $0.32 \, kg \, kg^{-1}$, a RMSE of $0.027 \, kg \, kg^{-1}$ was reported, but the data points still scattered strongly. The same effect was also reported by Demattê et al. (2006) who found the ratio of infrared to red wavelengths to decline with increasing soil moisture for a local set of soil samples. The generalisation of the models linking reflectance and moisture for a larger variety of soils seems difficult. It was postulated that using soil moisture tension instead of gravimetric or volumetric water contents could facilitate the generalisation of such models (Baumgardner et al., 1985, and references therein).

The effect of soil moisture on VNIRS estimates of other soil properties was investigated by Chang et al. (2005) using 190 sieved and crushed soil samples originating from one field. Separate PLSR calibrations for samples measured in moist and dried condition achieved comparable cross-validation performances. The assessed soil parameters included total, organic and inorganic carbon, total and mineralisable nitrogen, cation exchange capacity, and texture. All models failed to estimate samples of another field. Stenberg (2010) demonstrated that a standardised rewetting of soil samples can improve the estimation of certain soil properties. Significant decreases in RMSE of C_{org} and clay estimates were observed after rewetting soil samples to 0.2 and $0.3 \, m^3 \, m^{-3}$ volumetric water contents. The strong interactions between these soil constituents and water molecules may promote differences between soil samples and make them better 'visible' to the spectrometer.

5.2. Methods

5.2.1. Sample collection and preparation (wetting)

To assess soil moisture effects on spectroscopic measurements, 43 top soil samples (0-20 cm depth) were collected randomly in the surroundings of Bern, Switzerland (table A.1, page 119). The samples were collected without any predefined scheme to cover the region's soil variability. The samples consisting of one to two kilograms of soil were air dried and sieved (< 2 mm). A sub-sample of the air dried soil was poured into a petri dish and a VNIRS measurement was taken using a muglight (laboratory measurement setup described in section 2.3). The remaining soil was filled into a large aluminum shell. Deionised water was added while the soil material was simultaneously mixed until it was saturated. The samples were left for two days to equilibrate, then they were mixed thoroughly, a sub-sample was taken, filled into a petri dish, was compacted by hand, and measured by VNIRS (figure 5.3). Afterwards, the samples were air dried for three to

5. Soil moisture effects on VNIR reflectance

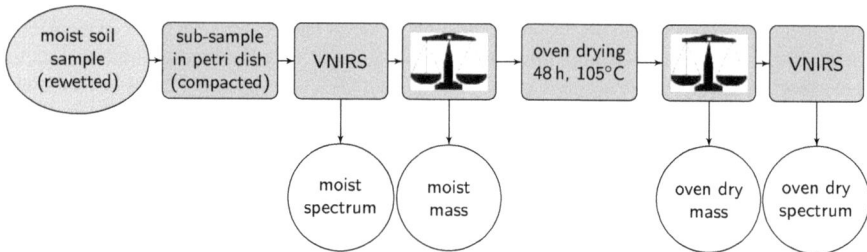

Figure 5.3 Flow chart: Measuring procedure for moist (and air dry) soil samples.

four days, before the next sub-sample was taken and measured. This procedure was continued until the samples were air dry again. In total, six sub-samples were measured for every sample from wet to air dried condition. All sub-samples (air dried and wet) remained in the petri dish for 48 h oven drying at 105°C to determine the gravimetric water content ω (equation 5.1). All sub-samples were remeasured in oven dry condition ensuring that the samples were not mixed or compressed between the moist and the oven dried status measurement.

$$\omega = \frac{\text{moist mass} - \text{oven dry mass}}{\text{oven dry mass}} \qquad (5.1)$$

5.2.2. Correcting for packing effects

As described above, a sub-sample of moist soil was filled into a petri dish and compressed by hand for the VNIRS measurements. After oven drying, the petri dishes were remeasured without any further mixing or compressing of the sample. These spectra represented the soil samples in oven dry condition, and therefore, the spectra of all sub-samples corresponding to the same sample were expected to be comparable. Representative of all samples, the spectra of sample nine are provided by figure 5.4. While the reflectance spectra of the moist sub-samples showed the expected differences related to the water content (figure 5.4 a), the spectra of the identical sub-samples after oven drying still differed (figure 5.4 b). The broad water bands near 1400 and 1900 nm disappeared and all spectra exhibited identical shapes, but the higher the former water content, the more the curve was shifted to lower reflectance values. The observed discrepancies were attributed to packing effects occurring for moist soil. Dried and sieved soil could easily be filled into the petri dishes, whereas the process got more and more complicated for wet soil samples. The moist soil samples possessed ductile characteristics impeding the compaction of the soil material within the petri dish. Thus, the wet samples were filled in less compactly into the petri dishes than dry samples, and therefore, less radiation was reflected. Because the effect was considered to be multiplicative (the less compacted the soil, the lower its reflectance), the measured spectrum for every sub-sample was corrected by taking the (oven dried status) spectrum of the corresponding sub-sample filled into the petri dish in air dried condition as reference (e.g. the red spectrum in figure 5.4 b for all measurements of sample nine). The correction was conducted for every sub-sample and wavelength λ by equation 5.2 in which $R(\lambda)$ represents the measured reflectance value, $R(\lambda)_{\text{oven dried}}$ the corresponding measurement of the same sub-sample after

5.2. Methods

Table 5.1 Assessed indicators for water content.

Single wavelengths	
1200 nm	No specific water absorption band
1420 nm	Strong water absorption band
1650 nm	No specific water absorption band
1780 nm	Weak water absorption band
1840 nm	At left edge of strong water absorption band
1940 nm	Strong water absorption band
2140 nm	At right edge of strong water absorption band
2200 nm	No specific water absorption band
Wavelength ratios	
2140:1840 nm	Indicator for inclination of spectrum
1940:1840 nm	Relative depth of water band to its right edge
1940:2140 nm	Relative depth of water band to its left edge
Relative areas	
1840-2140 nm	Relative area of water band
2140-2440 nm	Area indicating the spectrum's inclination

oven drying, $R(\lambda)_{\text{reference, oven dried}}$ the matching reference measurement, and finally $R(\lambda)_{\text{corr}}$ represents the corrected reflectance value for the moist sub-sample (provided in figure 5.4 c). The corrected spectra were used for all further calculations within chapter 5.

$$R(\lambda)_{\text{corr}} = R(\lambda) \cdot \frac{R(\lambda)_{\text{reference, oven dried}}}{R(\lambda)_{\text{oven dried}}} \tag{5.2}$$

5.2.3. Data analysis

Different parameters were derived from the corrected reflectance spectra including single wavelengths, ratios of wavelengths, and proportions of integrals (table 5.1 and figure 5.4 d-f). The included wavelengths were selected to cover ranges of the spectra influenced strongly by water content as well as ranges with minor influences. The wavelengths ratios (1940:1840 and 1940:2140 nm) and the relative area F_1 were designed to capture the extension of the water band around 1940 nm. The wavelength ratios related the maximum of the absorption band with its edges, while F_1 assessed the absorption area. The area under the reflectance curve between 1840 and 2140 nm was divided by the polygon limited by a straight line between the reflectance values at 1840 and 2140 nm (equation 5.3, where d_λ represents the interval in nm between the data points and R_λ the reflectance value at wavelength λ). As a result of its definition, F_1 values are decreasing for increasing size of the water absorption band. The same is true for the two wavelength ratios.

$$F_1 = \frac{\int_{1840\,\text{nm}}^{2140\,\text{nm}} R(\lambda)\,d\lambda}{(R_{1840\,\text{nm}} + R_{2140\,\text{nm}})/2 \cdot 300\,\text{nm}} \approx \frac{\sum_{1840\,\text{nm}}^{2140\,\text{nm}} R(\lambda) \cdot d_\lambda}{(R_{1840\,\text{nm}} + R_{2140\,\text{nm}})/2 \cdot 300\,\text{nm}} \tag{5.3}$$

The remaining wavelength ratio 2140:1840 nm was selected as an indicator of the general inclination of the spectra to the right hand side which reflects the influence of the fundamental vibrations

5. Soil moisture effects on VNIR reflectance

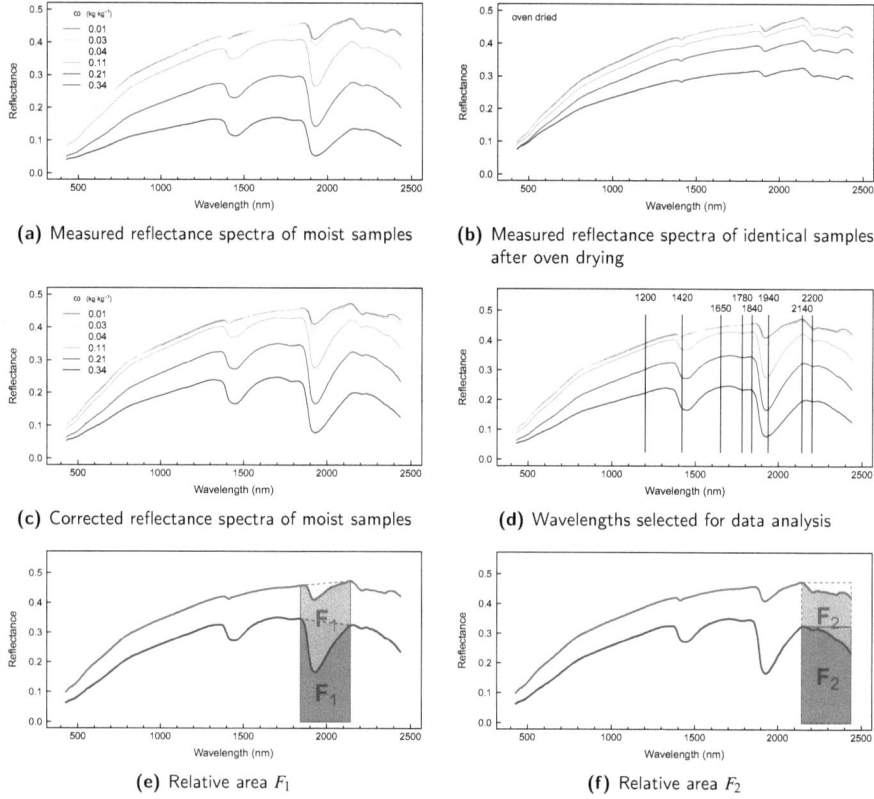

Figure 5.4 Reflectance spectra of (sieved) sample nine at different soil water contents ω (a-c), and illustration of the parameters deduced from the spectra for data analysis (d-f).

of water with their maxima in the MIR. The relative area F_2 (figure 5.4 f) was chosen with the same intention: the area under the reflectance curve between 2140 and 2440 nm was divided by the rectangle defined by the reflectance value at 2140 nm which equaled the maximum reflectance for this interval (equation 5.4).

$$F_2 = \frac{\int_{2140\,nm}^{2440\,nm} R(\lambda)\,d\lambda}{R_{2140\,nm} \cdot 300\,nm} \approx \frac{\sum_{2140\,nm}^{2440\,nm} R(\lambda) \cdot d_\lambda}{R_{2140\,nm} \cdot 300\,nm} \tag{5.4}$$

The deduced indicators were assessed qualitatively by plotting them for soil moisture gradients for single soil samples. Based on the visual inspection, the most promising (F_1 and the reflectance ratio 1940:1840 nm) were selected for subsequent regression analysis. Quadratic models were established to estimate soil moisture by the selected indicators. In addition, it was tested if the models were improved by including C_{tot} as auxiliary variable because it was concluded from the visual inspection that carbon influences strongly the relation between soil moisture and reflectance (cf. section 5.3.1). The models were built using the R function lm(). For the models including the auxiliary variable C_{tot}, insignificant terms were removed by applying the step() function with setting parameter k=4 which favoured smaller models over more extensive ones. All models were tested by six-fold cross-validation. The cross-validation segments were selected ensuring that all measurements of the same sample fell into the same group.

5.3. Results

5.3.1. Visual inspection of deduced parameters

The indicators deduced from the spectral data were analysed visually. Representative of all samples, the deduced indicators in relation to soil moisture of seven selected soil samples are shown (figure 5.5 for six indicators; figures A.20 and A.21, pages 129ff. for all indicators). Considering the assessed single wavelengths, almost similar behaviour was found for all of them irrespective of its location within or outside of water absorption bands, except for reflectance at 1940 nm. The latter exhibited an almost linear decrease in reflectance with increasing moisture. For some soil samples, the decrease in reflectance was slightly decelerated (figure 5.5 b). All other assessed wavelengths exhibited a clearly accelerated decrease in reflectance for increasing soil moisture (similar to figure 5.5 a). For low water contents, the wavelengths within the two water absorption bands (around 1450 and 1940 nm) and above 2000 nm exhibited a distinctive behaviour compared to the remaining wavelengths: the latter exhibited stable reflectance values for water contents from 0.00 up to 0.05 or 0.10 kg kg^{-1} (depending on the soil sample), whereas the first exhibited decreasing reflectance values over the whole moisture range. Generally, soil samples with lower reflectance in dry condition showed minor reflectance decreases when expressed as absolute value (as in the presented figures), but bigger decreases when expressed relative to the reflectance value in dry condition.

The wavelength ratios 1940:1840 and 1940:2140 nm as well as the relative area F_1 – all designed to assess the water absorption band around 1940 nm – showed similar behaviour (figures 5.5 c,e): the indicators decreased with increasing soil moisture, and the decrease was slightly decelerating for higher moisture contents. The slope and the curvature of the decrease differed between soil samples. The decrease seemed less steep and more curved for soil samples exhibiting elevated contents of C_{tot}, e.g. samples 5 and 38. Comparing the measurements of the air dried samples, the

5. Soil moisture effects on VNIR reflectance

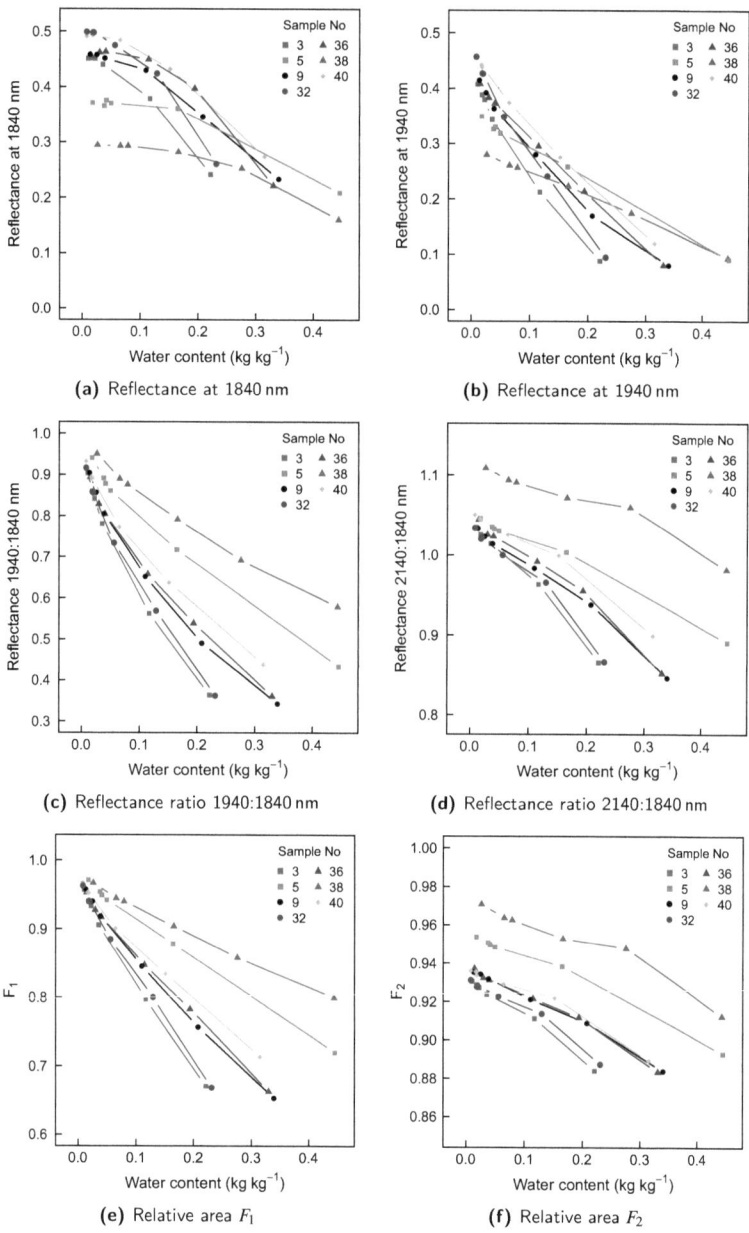

Figure 5.5 Influence of soil moisture on some of the selected indicators deduced from spectral data for seven randomly selected soil samples.

Table 5.2 Performance of different models to estimate soil water content ω ($n = 258$) by parameters deduced from reflectance data (relative area F_1 and r_λ = reflectance ratio 1940:1840 nm) and the auxiliary variable C_{tot}. R^2 and $RMSE_{Cal}$ were reported for calibration, $RMSE_{CV}$ for six-fold cross-validation.

Model equation $\omega =$	R^2	$RMSE_{Cal}$ — (kg kg^{-1}) —	$RMSE_{CV}$
$1.01 - 1.04 \cdot F_1$	0.81	0.054	0.054
$0.84 - 0.62 \cdot F_1 - 0.26 \cdot (F_1)^2$	0.81	0.054	0.054
$1.11 - 1.97 \cdot F_1 + 0.85 \cdot (F_1)^2 + 0.06 \cdot C_{tot} - 0.07 \cdot C_{tot} \cdot (F_1)^2$	0.92	0.034	0.037
$0.50 - 0.55 \cdot r_\lambda$	0.79	0.056	0.056
$0.61 - 0.94 \cdot r_\lambda + 0.31 \cdot (r_\lambda)^2$	0.80	0.055	0.056
$0.47 - 0.98 \cdot r_\lambda + 0.52 \cdot (r_\lambda)^2 + 0.05 \cdot C_{tot} - 0.05 \cdot C_{tot} \cdot r_\lambda$	0.91	0.037	0.040

values of these indicators were close with slightly higher values for the samples with high C_{tot} contents – these samples also exhibited higher water contents in air dry condition.

The wavelength ratio 2140:1840 nm and the relative area F_2 – designed to assess the inclination of the spectra to the right hand side – also showed decreasing values with increasing soil moisture, but the relation seemed to be a bit more complicated with a decelerating decrease at lower moisture contents followed by an accelerating decrease for higher moisture contents (figures 5.5 d,f). Some of the slopes were vertically shifted, especially for samples with high C_{tot} contents.

5.3.2. Regression models

The regression coefficients and the performance of the calculated models are provided in table 5.2, whereas the complete R output can be found in section A.3.1 (pages 132ff.). Using either F_1 or the reflectance ratio 1940:1840 nm (r_λ) as explaining variable, a lot of the variance in ω was explained ($R^2 > 0.8$, RMSE around 0.055 kg kg^{-1}). The performance of the models was greatly improved by including C_{tot} as auxiliary variable ($R^2 > 0.9$, RMSE between 0.034 and 0.040 kg kg^{-1}). The models exhibited a significant linear term for C_{tot} implying a positive offset for soil samples with higher C_{tot} contents. Furthermore, they exhibited so-called interaction terms $-C_{tot} \cdot (F_1)^2$ and $C_{tot} \cdot r_\lambda$ – indicating changes in the relation ω to F_1 and r_λ, respectively, induced by differences in C_{tot}. Generally, the models using the relative area F_1 performed better than the corresponding models using the reflectance ratio r_λ. The models' performance were very similar for calibration and six-fold cross-validation.

5.4. Discussion

First, the methodological problems preparing moist soil samples for spectroscopic measurements will be discussed. As explained in section 5.2.2 and shown in figure 5.4 a-c, it was impossible to fill in moist soil material into petri dishes as compactly as dry soil, because the added water increased the material's cohesion. Therefore, the reflectance for the whole wavelength range of the rewetted soil samples was clearly reduced, even after oven drying. In other words, the albedo was reduced because less soil per area was measured. Generally, the wetter the sample when it

5. Soil moisture effects on VNIR reflectance

Table 5.3 Soil samples used for figure 5.5. C_{tot} and N_{tot} determined by CN analyser, *pH* by Hellige pH indicator.

Sample	Origin	Land use	pH	C_{tot} (g kg^{-1})	N_{tot} (g kg^{-1})
3	Wabern	pasture	6.5	26	2.5
5	Kehrsatz	forest	4.5	77	5.0
9	Kehrsatz	crop rotation	7	57	4.2
32	Gurten	pasture	4.0	20	2.3
36	Kerzers	crop rotation	7	52	4.8
38	Ins	crop rotation	7	128	8.9
40	Brienz	crop rotation	6	28	3.1

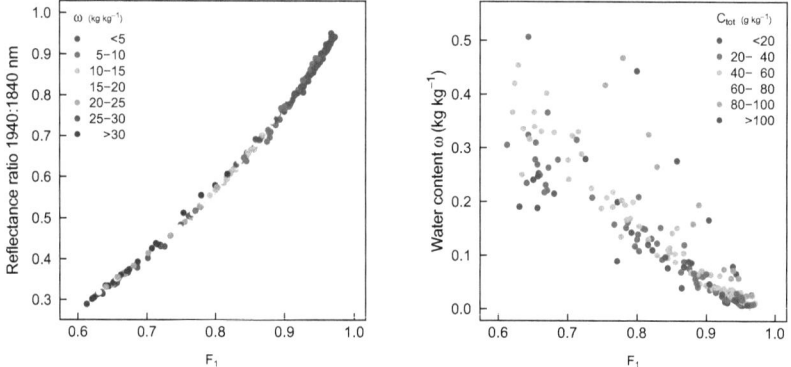

Figure 5.6 Correlation between water content ω, F_1, and reflectance ratio 1940:1840 nm for all soil samples ($n = 258$). Colours indicate water content and C_{tot} classes, respectively.

70

5.4. Discussion

was filled in, the bigger the albedo reduction. It was assumed that the effect on the spectra for moist and oven dried samples was the same. Because all samples were oven dried and remeasured then, the original spectra could be corrected for the error introduced by the described packing effects. The procedure described in section 5.2.2 was considered to eliminate the reduction in albedo successfully, although a reduction of the signal-to-noise ratio must be expected. Anyhow, the methodology should be improved to avoid such packing effects. One possibility: fill in dry soil samples compactly into a box and wet it without mixing. The problem with this procedure: moisture may differ at the surface and the interior of the sample, and it seems difficult to determine the relevant water content.

The observed effects of soil moisture on reflectance (figures 5.4 c and A.19, pages 120ff.) were expected and consistent with published works like Bowers & Hanks (1965), Baumgardner et al. (1985), and Stenberg et al. (2010). These effects – namely albedo reduction, bigger and broader water bands around 1400 and 1900 nm, inclination of the spectral curve to the right hand side – were explained by the spectroscopic mechanisms outlined in section 5.1. When considering the reflectance spectra (left column of figure A.19), the effects seemed to be very similar for all samples and wavelengths. Considering the relative changes in reflectance (the reflectance of the moist sample was divided by the reflectance of the dry sample, right column in figure A.19), differences were better accessible. The most remarkable feature was that for wavelengths below 600 to 700 nm, the reduction in reflectance was smaller compared to the rest of the spectrum for high water contents. For some soil samples – among others, samples 5 and 19 – the reflection in this area was bigger for high water contents than for the dry samples. It was supposed that for some samples at high moisture levels, excess water was present at its surface increasing the reflectance strongly. The thickening water films altered the refractive index of the measured surface and increased the diffuse reflection of the shorter wavelengths.

5.4.1. Deduced parameters

Different parameters were deduced from the spectral data to assess the changes due to soil moisture. The selected single wavelengths (1200, 1420, 1650, 1780, 1840, 1940, 2140, and 2200 nm; figures 5.3.1 a-b and A.20) exhibited decreasing reflectance with increasing moisture contents. The decrease in reflectance was accelerating, except for 1940 nm where stable and decelerating decreases were observed. These observations are in contradiction to the findings of Bowers & Hanks (1965) and Skidmore et al. (1975) who found clearly decelerating decreases for all wavelengths. The accelerating decreases observed in our experiments were certainly not caused by the applied correction for packing errors discussed above. Before correction, reflectance values were lower, and thus the acceleration of the decrease even stronger. The range of water contents was bigger for this work (0 to 0.4 kg kg^{-1}) compared to the cited studies (0 to 0.2 or 0.25 kg kg^{-1}), but considering a narrowed range of water contents did not change our findings.

Surprisingly, no striking differences were found between wavelengths situated outside and inside of the observed water absorption bands, with 1940 nm being the only exception. The other assessed wavelengths showed very similar behaviour, except that the range of reflectance values was slightly bigger for those situated inside water bands. This observation was interpreted that the albedo reduction observed for the whole spectra was the dominant process for these wavelengths. Reflectance at 1940 nm showed clearly distinct behaviour indicating that absorption processes related to the water band situated there strongly influenced reflectance. This emphasised the goodness of this part of the spectra to predict soil moisture. Nevertheless, it was concluded that single wavelengths were not very useful to estimate soil moisture contents because the observed curves changed considerably between soil samples, and because the starting points (the data points for the dry soil samples) of the curves deviated strongly. The latter problem could be resolved by calculating relative reflectance values (relative to dry condition), but the need to dry and remeasure the soil samples would make the measurements on moist samples obsolete.

Another interesting detail: for wavelengths below 1840 nm and outside the absorption band around 1450 nm, the reflectance values remained quite stable for low soil water contents. Depending on the soil sample, reflectance remained stable from 0.00 up to 0.05 or even 0.10 kg kg^{-1} (figure A.20). In contrast, reflectance started to decrease at very low water contents for wavelengths within the water bands around 1450 and 1940 nm and above 2000 nm. These findings imply that the latter wavelengths should be removed from VNIRS calibration models if they are dispensable for the estimation of the addressed soil property (meaning that the remaining wavelengths contain enough information). This could produce VNIRS calibrations much more stable to minor changes in soil moisture caused by varying air humidity. An example demonstrating the success of the suggested wavelength removal will be discussed in section 6.4.2 (page 92). Of course, there is a trade-off between calibration stability gained and spectral information lost by wavelength removal.

Addressing the calculated wavelength ratios and relative areas (figures 5.3.1 c-f), the parameters assessing the 1940 nm absorption band (F_1 and ratio 1940:1840 nm) were judged to be superior to the remaining to estimate soil moisture. F_2 and the reflectance ratio 2140:1840 nm both reflected well the spectra's inclination to the right hand side and both obviously correlated with soil moisture, but on one hand, the relation between soil moisture and these parameters seemed to be more complicated, and on the other hand, the relation differed more strongly between soil samples. In contrast, the relations between soil moisture and F_1 and the corresponding wavelength ratios seemed to be more regular, and their starting points (the data points for the dry soil samples) were quite close. Based on this considerations, F_1 and the ratio 1940:1840 nm were chosen for the subsequent regression analysis.

5.4.2. Regression models

Using the selected parameters F_1 and r_λ (representing the wavelength ratio 1940:1840 nm) together with C_{tot} as auxiliary variable, soil moisture contents were estimated successfully (table 5.2). The performance of the models using F_1 and r_λ were similar with F_1 being slightly superior. The two parameters were strongly correlated (figure 5.6 a), but F_1 may have been less prone to minor spectral changes as it included a spectral range and not only two wavelengths. Therefore, F_1 should be favoured to derive calibrations to estimate soil moisture.

Including C_{tot} into the models was required because the relation between the chosen parameters and soil moisture changed with varying C_{tot} contents. Organic matter exhibits numerous functional groups like hydroxyl, carboxyl, and phenolic groups (Gisi et al., 1997). Water molecules interact with these groups. Therefore, the presence of organic matter influences how and how strong water is captured by soil and also how it is distributed. For example, air dried soil samples with high organic contents exhibit higher water contents compared to air dried samples containing little organic matter (visible in figure 5.5: starting points representing air dry condition of samples 5 and 38 are shifted to higher water contents). Taking this into account, it seemed logical that the relation between soil reflectance and moisture also depended on the organic matter present. As demonstrated with the regression models, C_{tot} can be used to correct the VNIRS moisture content estimates for the amount of organic matter, although it remains to be tested if C_{org} would be even better for this purpose. C_{tot} captures organic carbon as well as inorganic carbon. The latter (most often present as calcium carbonate and its residues) exhibits carboxyl groups – as does organic matter – interacting with water, and therefore, C_{tot} may be the better choice. The composition of organic matter is highly variable depending on the environmental conditions (Gisi et al., 1997), thus it was assumed that also the quality of organic matter influenced the relation between soil moisture and reflectance.

Other soil constituents interacting strongly with water molecules are clay minerals (Clark, 1999). I am very optimistic that clay content could explain most of the unexplained variation in the presented regression models, although it was impossible to prove this hypothesis, because clay contents were not determined for the used samples. Stenberg (2010) demonstrated that it is even possible to take advantage of the circumstance that some soil constituents interact with water. He observed significant decreases in RMSE of C_{org} and clay estimates after rewetting soil samples to 0.2 and 0.3 $m^3 \, m^{-3}$ volumetric water contents.

In summary, it seems possible to estimate water contents of soil samples by VNIRS calculating the presented parameter F_1 from its reflectance spectra if their C_{tot} contents are known. Of course, further investigations are needed to analyse whether the found relations can be maintained for larger and more diverse sets of soil samples. Considering the implications on VNIRS field measurements, the presented regression models were bad news. They led to the conclusion that the effects of soil moisture and organic carbon (and presumably also clay content) cannot be distinguished in unknown spectra. Either, one of the parameters must be known, or it must be controlled. Consequently, field measurements must be conducted on fields where differences in soil moisture are minor, so that the used calibration models can be adapted to the actual soil moisture. Or yet another possibility would be to measure the moisture of the soil surface simultaneously by another technique applicable as on-the-go sensor.

5.5. Summary and Conclusions

Increased soil moisture has strong effects on soil reflectance: the albedo is reduced, and strong absorption bands around 1450 and 1940 nm are promoted. As these effects interfere with the estimation of other soil parameters like C_{org}, it is a prerequisite for VNIRS field measurements to know the relation between soil moisture and reflection. Within this project, reflectance of 43 soil samples originating from the region of Bern were measured at varying water contents from oven dry to saturated. For moist soil samples, errors due to packing effects were observed: the cohesion of the moist soil material inhibited its compaction leading to lowered reflection values. Replicate measurements after oven drying of the soil samples were used to correct for these errors.

Reflectance at single wavelengths as well as reflectance ratios and relative ares (F_1 and F_2) were assessed for their suitability to estimate soil water contents. Reflectance decreased for all wavelengths with increasing soil moisture. For wavelengths below 1940 nm situated outside the water band around 1450 nm, reflectance remained stable at low water contents (0.00 up to 0.05 or 0.10 $kg \, kg^{-1}$), whereas reflection started to decrease immediately for the remaining wavelengths. Parameters capturing the absorption band around 1940 nm seemed most promising to estimate soil moisture. Regression models were successfully calculated using the relative area F_1 or the reflectance ratio 1940:1840 nm in conjunction with C_{tot} as auxiliary variable. The model using F_1 was slightly superior to the remaining with a RMSE of 0.034 kg kg^{-1} and R^2 of 0.92. The great impact of organic matter was explained by the circumstance that its functional groups strongly interact with water. It was assumed that clay content could explain most of the unexplained variance in the models, as clay minerals interact with water, too.

Based on the presented results, it was concluded:

- Moist soil material cannot be filled in into petri dishes as compactly as dry soil. Consequently, packing effects are observed.

- The albedo reduction for VNIRS due to these packing effects can be corrected for by oven drying and remeasuring the samples.

- For certain soil parameters, more stable calibration models may be produced by discarding wavelengths around 1450 and 1940 nm and above 2000 nm.

- The relative area F_1 defined within this chapter was found to be most suitable to estimate soil moisture contents.

- The relation between F_1 and soil moisture is influenced by the soil's carbon content. Thus, it must be considered for calibrations.

- Further investigations are needed to test whether the found relations apply for more diverse sets of soils.

- Additionally, including clay content into the models should be assessed.

6. Field measurements

Since portable VNIR spectrometers exist, soil scientists dream to measure soil properties directly in the field (*in situ*; sometimes the term *proximal sensing* is used). This would allow for faster and cheaper analyses compared to laboratory spectroscopy because collecting, drying and sieving soil samples would become obsolete. In particular, stakeholders interested in precision farming methods expect that VNIRS in situ measurements will enable the determination of soil properties like nutrition status at high resolutions and, thus, allow more efficient use of resources like fertilisers and agro-chemicals resulting in economical and ecological benefits (Hummel et al., 1996; Maleki et al., 2007; Mouazen et al., 2007b). Fast and comprehensive measurements of whole fields or even regions are also a promising perspective for soil mapping and monitoring, e.g. to estimate soil carbon stocks and their changes (Stevens et al., 2006). Furthermore, there are attempts to link proximal sensing measurements with remote sensing data collected by air-borne devices and satellites (Stevens et al., 2008).

Depending on the specific needs, different measuring setups are thinkable:

- **Contact measurements** of the intact soil using a contact probe with an internal light source. Can be used to measure the soil surface, but also subsoil fractions can be accessed, either by digging holes and soil profiles, or by extracting soil cores.

- **Short distance measurements** of the soil surface using either a hand held probe or a probe fixed to a construction. The measurements are conducted either with sun light or an artificial light source.

- **On-the-go sensors** are able to measure soil reflectance while moving and can be constructed as short distance or contact measurement sensors. Depending on the construction, soil surface or subsoil measurements are collected. An artificial light source is indispensable.

Regardless of the chosen measuring setup, VNIRS in situ measurements are influenced by the field conditions. Soil moisture seems to be one of the dominant factors (Baumgardner et al., 1985, and chapter 5 of the present work), but also soil surface roughness, the presence of stones and plant residues, and contamination of the VNIRS probe by dust or smearing can have major impacts on the measurements (Stenberg et al., 2010). Furthermore, when not using a contact probe, changes in the distance from the sensor to the soil and also changes in solar irradiation can corrupt the measurements (Sudduth & Hummel, 1993b). Therefore, spectral data of laboratory and field measurements are not comparable directly. Either separate calibrations, or mathematical pretreatments to make laboratory and field data comparable are needed.

The present work was part of the iSOIL project that focused on improving fast and reliable mapping of soil properties, soil functions and soil degradation threats. Within the project framework, a portable VNIRS device was integrated on a mobile platform to conduct VNIRS measurements directly in the field. Section 6.1 provides an overview of published studies on VNIRS field measurements and existing on-the-go sensors. The platform used for this project and measuring procedures are described in section 6.2, the results of the field measurements are presented and evaluated through sections 6.3 and 6.4.

6. Field measurements

6.1. State of the art

6.1.1. In situ measurements

Almost 20 years ago, Sudduth & Hummel (1993a) presented a prototype portable near infrared spectrophotometer with a sensing range from 1630 to 2650 nm wavelengths, installed it to a vehicle, and tested it for in situ soil organic matter sensing (Sudduth & Hummel, 1993b). The surface of a 3.5 to 5 cm flat-bottomed furrow prepared by some type of opener was measured underneath an enclosure (to keep away environmental light) at a speed of 0.65 m s^{-1}. The test fields at Urbana, IL, USA, exhibited C_{org} contents from 14 to 34 g kg^{-1} (mean $\pm$ sd: $24 \pm 5 \text{ g kg}^{-1}$). Laboratory measurements of 30 independent soil samples (re)wetted to two different moisture tensions (1.5 and 0.033 MPa, field capacity and permanent wilting point, respectively) served for calibration by PLSR. Although a good cross-validation RMSE of 2.8 g C kg^{-1} was achieved for the calibration, C_{org} in situ estimations showed neither good agreement, nor good correlation with corresponding laboratory measurements resulting in a RMSE of 5.3 g C kg^{-1}. The authors theorized that variations in sensor-to-soil distance or vibrations of the moving sensor introduced these errors.

Laboratory and field spectroscopy were compared by Udelhoven et al. (2003) using 114 in situ measurements and corresponding soil samples (0 to 1 cm soil depth) from an agricultural plot in the Trier region, Germany. For in situ measurements, they used an ASD FieldSpec II and a reflectance probe equipped with an external illumination source and exhibiting a field of view of 10 cm radius. Prior to the measurements, the soil surface was leveled. Although they were successful in estimating concentrations of organic and inorganic carbon, nitrogen, iron, calcium and plant available magnesium based on laboratory spectroscopy, the calibrations based on in situ measurements performed inferior and were considered of no practical use by the authors. Unfortunately, no RMSE were reported, but the conclusion were supported by the reported coefficients of correlation (R^2) that were considerably lower for in situ compared to laboratory spectroscopy (e.g. 0.56 and 0.84 for Fe estimations in situ and at the laboratory, respectively). The large errors were mainly attributed to the measuring setup and the small field of view of the probe, but of course also differences in water content cannot be neglected.

Waiser et al. (2007) collected 72 soil cores to a maximum depth of 105 cm from six fields in Texas, USA, and measured in situ soil reflectance of 270 sub-samples using a contact probe immediately after halving the cores lengthwise. By using 70 % of the cores for calibration and the remaining for validation, they achieved a RMSE of $61 \text{ g clay kg}^{-1}$ and where able to estimate the clay distribution for the soil profiles. When measuring the intact cores after driying, RMSE decreased to 41 g kg^{-1}, whereas for dried and ground soil samples, a RMSE of 62 g kg^{-1} was reported. Smearing the moist soil cores deliberately to simulate the possible effect of inserting a reflectance probe into the soil increased RMSE to 74 g kg^{-1}. Higher RMSE and partly very big bias resulted when leaving out all samples from entire fields for validation. Also Viscarra Rossel et al. (2009) tried to estimate clay content using in situ measurements. The reflectance measurements were taken from ten soil profiles using a contact probe. An existing set of over 1200 spectra measured under laboratory conditions were spiked with 74 in situ measurements from soil profiles. Absorbance spectra were transformed by continuum removal and wavelength regions dominated by water were discarded to minimise differences between laboratory and field measurements. The performance of the VNIRS clay estimations differed strongly between soil profiles: while the clay distribution could be estimated reasonably for some, poor results were reported for others. Surprisingly, estimations for the in situ measurements performed slightly better than corresponding laboratory measurements with validation RMSE of 73 and 83 g kg^{-1}, respectively. Possibly, differences in clay content were better detectable under in situ conditions because soil moisture was strongly influenced by clay content. Recent results by Stenberg (2010) support this argumentation. Another explanation could be the fact that in situ spectra of the profiles were included for calibration, but not the corresponding laboratory measurements.

An accessory for ASD spectrometers to collect sub-surface reflectance measurements in drilling holes was constructed by Ben-Dor et al. (2008). Soil profiles down to one meter depth were measured in increments of 10 cm. The spectral data was used to describe the soil layers and to assess soil colour, soil moisture, the specific surface area, organic matter contents, soil carbonates, and free iron oxides. Except for free iron oxides, good results were reported for all of the targeted soil properties allowing to estimate these soil properties over the whole soil profile. It needs to be mentioned that the PLS regression was restricted to only four selected soil profiles and, thus, was very specific for the validation samples. Furthermore, it remained obscure if the presented VNIRS estimates of soil properties for the profiles – that were really appealing – originated from the calibration, cross-validation or validation step.

Stevens et al. (2008) measured soil surface reflectance of bare Belgian fields under environmental illumination from a distance of one meter with an ASD FieldSpec Pro. For each plot, a representative spectrum (average of nine measurements capturing a field of view of 0.45 m diameter) was compared with the C_{org} content of the corresponding mixing sample collected from the top 5 cm soil layer. They removed spectra influenced by vegetation, eliminated wavelengths related strongly to water vapour (1340 - 1430, 1810 - 1970, 2400 - 2500 nm), and tested various combinations of smoothing and derivation algorithms to find the optimum data pre-treatment for subsequent PLS regression. For this set consisting of 100 soil samples (C_{org} mean $\pm$ sd: 13.4 ± 2.7 g kg^{-1}), a leave-out-one cross-validation RMSE of 1.2 g C kg^{-1} was achieved – equal to the RMSE achieved for laboratory measurements of the same samples dried and sieved (< 2 mm). Applying the calibration on other field measurements from the same region taken two years earlier resulted in bad estimations, thus Stevens et al. (2008) concluded that 'a calibration is still needed before each measurement campaign.' Nevertheless, the study showed that under good conditions and appropriate data pre-treatment field spectroscopy can be competitive to laboratory spectroscopy.[1]

The presented results show that in situ measurements are promising, but useful results were only reported for small, specific sets of measurements. For most studies, portable and rugged VNIR spectrometers of ASD were used with different optical interfaces. All authors used spectral pre-treatments (first and second derivatives, smoothing algorithms and multiple scatter correction, and combinations of them) and subsequent PLSR calibration, except Viscarra Rossel et al. (2009) who used continuum removal and bagging-PLSR. Additionally, Stevens et al. (2008) and Viscarra Rossel et al. (2009) removed wavelengths related to water.

6.1.2. On-the-go sensors

During the last two decades, there has been a lot of research and technical innovation in the area of soil spectroscopy, especially for VNIRS on-the-go sensors. For few years, such sensors are commercially available, e.g. from Veris Technologies (described in Christy, 2008). The sensor can be mounted to a tractor with an adjustable sampling soil depth from 2.5 to 10 cm (Figure 6.1).

---
[1] In contrast, measuring the same fields by an air-borne VNIR device resulted in unsatisfactorily C_{org} estimates and clearly higher RMSE (1.7 g C kg^{-1}).

6. Field measurements

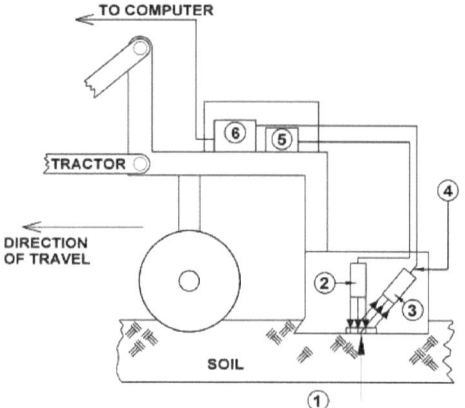

Figure 6.1 Shank-based spectrophotometer used by Christy (2008) to obtain NIR spectra. (1) Sapphire window; (2) halogen lamp; (3) collection optic; (4) fiber optic; (5) spectrometer; (6) power supply. Illustration by Christy (2008).

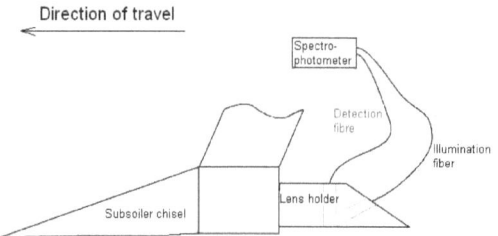

Figure 6.2 Schematic illustration of the NIR spectrophotometer sensor used for on-the-go measurements presented by Mouazen et al. (2005a, illustration therein)

Spectra are collected in the range from 350 to 2200 nm wavelengths. The included software can be used to select the locations for reference sampling. Christy (2008) analysed eight fields for soil organic carbon and took 15 reference samples per field that were analysed by conventional analytical methods. They achieved a RMSE of $5\,\mathrm{g\,C\,kg^{-1}}$ in one field-out cross-validations (the values of one field were estimated using the samples of the seven remaining fields for calibration). It is important to note that all fields belonged to a small area and, thus, must be considered as a relatively large local calibration (more than 100 samples).

Also Mouazen et al. (2005a) developed an on-the-go sensor. It consists of a subsoiler penetrating the soil. The reflectance is measured in the trench built by the subsoiler (see Figure 6.2). The sensor was tested to measure soil moisture (Mouazen et al., 2005a, 2006), total and organic carbon (Mouazen et al., 2007b), phosphorus (Maleki et al., 2007, 2008), soil texture classes (Mouazen et al., 2005b), and soil colour groups (Mouazen et al., 2007a). The results of these investigations were promising although the maps generated by the VNIRS sensor were not matching satisfactorily the maps generated by the reference methods (Mouazen et al., 2007b). For C_{tot} maps, a relative mean error of 6 % was reported. Changes in the sensor-to-soil distance when crossing tramlines or other bumps were considered as major problem.

Table 6.1 Specifications of the mobile platform constructed for the present work.

Length (without/with cover)	100/150 cm
Width	35 cm
Height	57 cm
Weight (platform only)	6 kg
Weight (platform fully equipped)	circa 30 kg
Sensor-to-soil distance	14 cm
Sensor field of view	circular area, 5.7 cm diameter
Illumination	Halogen lamp, 100 W, 12 V D.C. 3200°K colour temperature
Operation time	3 - 4 hours per battery charge

6.2. Methods

6.2.1. Mobile platform

Within the framework of the present work, a mobile platform to perform VNIRS measurements directly in the field was constructed. The platform was equipped with a FieldSpec 3 spectrometer (ASD Inc., Boulder, CO, USA), a halogen headlamp (Lowel i-100; Lowel-Light Mfg, Hauppauge, NY, USA) for illumination, the spectrometer's genuine accumulator, and a car battery supplying the headlamp. The headlamp and the VNIRS probe were mounted at the rear of the platform. The probe was measuring reflectance perpendicular to the soil surface (see table 6.1 and figure 6.3). A cover prevented sunlight from interfering with the measurements, hence the measurements were not constrained to good and stable solar irradiation. The platform was moved manually. Its use was restricted to low velocities (one to two meters per second) to avoid damages to the equipment. The platform was used to conduct stop-and-go and continuous (on-the-go) measurements. Measuring intervals and collection time (the number of co-added scans) were adjustable by the spectrometer's software. For the white reference measurements, a 50 cm · 50 cm reflection target (Zenith Alucore, SphereOptics, Uhldingen, Germany) was placed underneath the cover.

6.2.2. Field site

The presented field measurements were conducted at the long-term fertilisation experiment Bad Lauchstädt (near Halle, Saxony-Anhalt, Germany) established in 1902. The field is flat (115 - 120 m above Sea level) and characterised as Haplic Chernozem with contents of 21, 67 and 12 % of clay, silt and sand, respectively (Leinweber et al., 1994). Yearly precipitation is 483.3 mm and the average temperature 8.8 °C. The field is divided into eight plots of 25.5 to 28.5 m width and 200 m length cultivated with a four-yearly rotation system (potato, winter wheat, sugar beet and barley). Each plot is further divided into 18 subplots of 10 m length for which different levels and combinations of manure and N, P, and K fertilisers are applied (figure 6.4). Because of the varying fertiliser and manure inputs, the contents of C_{org} and N_{tot} are differing strongly between the subplots with lower contents of C_{org} and N_{tot} and higher C/N ratios for the subplots with lower inputs (figure 6.5).

At the time of the VNIRS field measurements on April 7, 2010, plot seven exhibited a dry soil surface, no rain had precipitated the days before. Two weeks earlier, sugar beets had been drilled

6. Field measurements

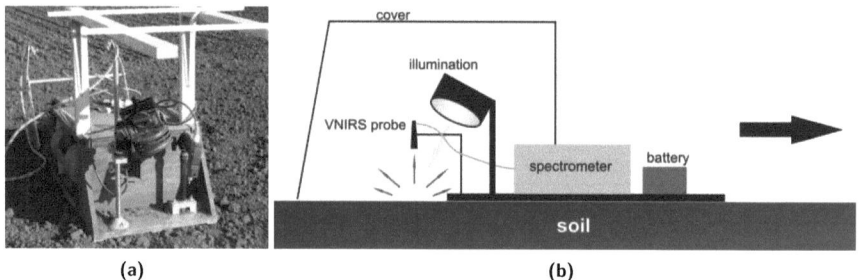

Figure 6.3 Setup used by the University of Bern for VNIRS field measurements by an ASD FieldSpec 3. (a) Rear end (cover removed) with illumination (left) and VNIRS probe (right). (b) Schematic of side view.

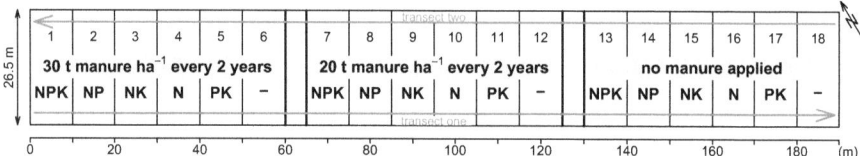

Figure 6.4 Long-term fertilisation experiment Bad Lauchstädt, plot seven: combinations of N, P, and K fertilisers and levels of manure applied to subplots 1 to 18. The green arrows represent the two transects measured within the present work.

6.2. Methods

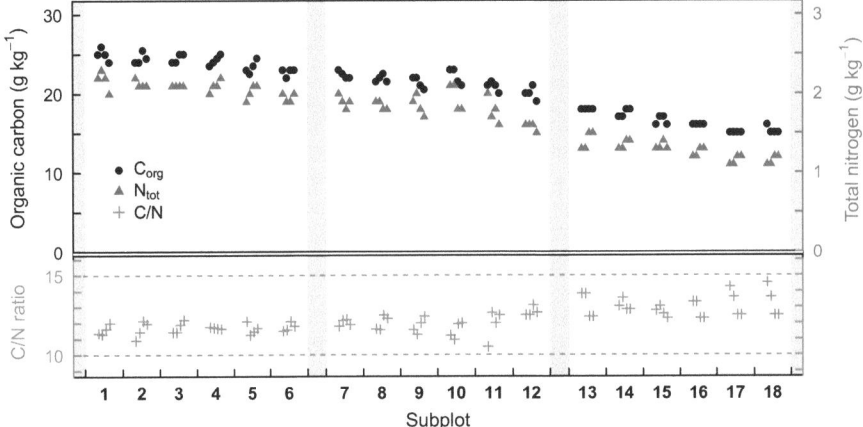

Figure 6.5 Top soil (0-20 cm) concentrations of C_{org} and N_{tot} as well as C/N ratios determined by dry combustion for plot seven (four replicates per subplot, samples collected in autumn 2009; data provided by Helmholtz Centre for Environmental Research – UFZ Leipzig).

with subsequent harrowing of the seedbed, thus the soil surface was free of plant residuals and smooth, with no visual differences in soil structure at the surface.

6.2.3. Field measurements

The field measurements were conducted on April 7, 2010 under good weather conditions with no clouds and moderate wind that gained strength during the day. Two transects covering the whole length of the plot were measured: transect one in eastward direction at a distance of about three meters from the southern border of the plot, and afterwards transect two in westward direction at a distance of three meters from the northern border (figure 6.4). The transects were measured in segments of 30 m in stop-and-go mode as well as on-the-go (except subplots one to nine of transect two where only stop-and-go measurements were taken due to low batteries).

Spectrometer and lamp were warmed up for 20 minutes. The two sets of batteries used allowed measuring during seven hours. For the **on-the-go** measurements, the spectrometer was set to co-add five scans to one spectrum and to save a spectrum every two seconds. The platform was moved at a speed of $0.5\,\mathrm{m\,s^{-1}}$, thus it took 60 s to measure one segment of 30 m resulting in 30 spectra representing roughly one meter each. Every segment was measured two times (the second time in the opposite direction). A white reference measurement was taken immediately prior to every run. The spectra were localised on the transect according to the time it was taken. During some measurements, the collection of some spectra was delayed, in this case the measurement of the segment was repeated. For the **stop-and-go** measurements, one spectrum based on 30 co-added scans was taken every two meters. Because the platform was stopped for the measurements, the spectra could be localised more accurately than the on-the-go spectra. A white reference

6. Field measurements

measurement was taken immediately prior to measuring each segment.

6.2.4. Laboratory measurements

For every subplot, two soil samples were collected on each of the transects at four and six meters distance from the border of the subplot at exactly the same locations where stop-and-go measurements were taken. The soil samples were collected from the top five centimeters and sealed in plastic bags to conserve the soil moisture. At the laboratory, the soil water content was determined gravimetrically by 48 h oven drying. The dried soil samples were sieved ($< 2\,\text{mm}$) and VNIR spectra were taken using the same FieldSpec 3 as in the field together with a muglight contact probe. Nine randomly selected samples from transect one were also analysed for C_{tot} and N_{tot} by dry combustion (CN analyser) for validation purposes.

For calibration, 34 soil samples (0 - 20 cm) collected by UFZ Leipzig-Halle in October 2008 at plot two of the long-term experiment were used. The set consisted of each six samples of subplots 1, 6, 7, 12, 13, and 18. The sieved and dried samples were measured by VNIRS. Additionally, C_{tot} and N_{tot} contents were measured by dry combustion (Vario EL III Element analyser, Elementar Analysensysteme GmbH, Hanau, Germany, cf. section 2.4.4). Because no inorganic carbon was present in the samples, C_{tot} was assumed to equal C_{org}.

6.2.5. Data analysis

For all used spectra, the amount of data was reduced by using only the reflectance at every fourth wavelength. At first, the spectra of the field and laboratory measurements were screened visually to assess the quality of the spectra and the differences between the measuring setups. For calibration, different spectral pre-treatments were alternatively applied to the collected reflectance (R) spectra and compared:

- Transformation to absorbance ($A = -\log R$)

- First derivative of absorbance using a Savitzky-Golay filter of 25 nm length (Savitzky & Golay, 1964)

- Same pre-treatments in combination with subsequent removal of wavelengths effected strongly by water content (1340 -1430, 1810 -1970, 2400 -2500 nm)

Based on 34 calibration samples from plot two, PLS models to estimate C_{org} and N_{tot} were established for the data generated by the different pre-treatment techniques. One sample was excluded because there were hints that its spectrum was not measured correctly. The optimum number of PLS factors was assessed by the RMSE of six-fold cross-validation. The resulting models were applied to the spectral data collected at the laboratory and the field for plot seven.

Furthermore, calibration models based on augmented collections of soil samples were built by adding the spectral data of the nine validation soil samples collected from transect one for which C_{org} and N_{tot} contents were known. Selecting few samples from the ensemble of samples that one is willing to analyse and adding these to the calibration set is usually called *spiking*. First, the 33 calibration samples were spiked with the spectra collected in the field by stop-and-go corresponding to the nine samples. The term *spiking with field data* will be used for this models. Second, the calibration samples were spiked with both the spectra collected in the field and the laboratory corresponding to the nine samples. The term *spiking with field and lab data* will be used for this models. The spiked models were only calculated for the spectral pre-treatment that performed best for the calibration sets that were not spiked.

Finally, calibration models using only the stop-and-go field spectra corresponding to the nine validation samples from transect one were calculated and applied to the field measurements. This procedure was of interest because it was the simplest and least laborious – and consequently less expensive – calibration model and, usually, if a new, unknown field is analysed, there are no archived and analysed soil samples that can be used for calibration.

To validate the C_{org} and N_{tot} estimates calculated for the laboratory VNIRS measurements, data provided by Helmholtz Centre for Environmental Research – UFZ Leipzig were used. They analysed (dry combustion by CN analyser) four soil samples (0-20 cm depth) per subplot collected at plot seven in autumn 2009. To validate the C_{org} and N_{tot} estimates calculated for the field measurements, the previously calculated estimates for the laboratory VNIRS measurements were used because they were assumed to be accurate and more appropriate to asses the collected field data (cf. subsection 6.4.4).

6.3. Results

6.3.1. Visual screening

The reflectance data corresponding to plot seven collected at the laboratory and in the field by stop-and-go and on-the-go measurements exhibited clearly visible differences. In Figure 6.6, the data corresponding to the samples taken from transect one within subplot two were compared for illustration. The measurements from the remaining subplots showed similar behaviour. Reflectance values measured at the laboratory (figure 6.6 a) were substantially higher than those measured at the field (6.6 b,c), Furthermore, the shapes differed – especially for the wavelengths influenced strongly by water content (blue areas in figure 6.6). Certainly, the air dried samples measured at the laboratory exhibited lower water contents compared to the intact soil surface measured in the field. The gravimetric water content determined for the soil samples (0-5 cm depth) collected from the two transects ranged from 0.07 to 0.19 kg kg^{-1}. The field measurements also seemed to be more noisy, their lines were less smooth compared to laboratory measurements. Additionally, the discontinuities at the sensor changes were more pronounced for the field measurements (red arrows in figure 6.6). When comparing stop-and-go with on-the-go measurements, the first exhibited much more differences between the spectra: while the shape of the spectra seemed to be quite similar for all spectra, the whole spectra were shifted to higher or lower reflectance values for the stop-and-go measurements, whereas much smaller shifts were observed for the on-the-go spectra.

6.3.2. Calibration

No striking differences were apparent when comparing cross-validation RMSE of the calibrations based on the four tested spectral pre-treatments. Using the optimum number of PLS factors, cross-validation RMSE from 1.2 to 1.5 g C kg^{-1} and from 0.13 to 0.15 g N kg^{-1} were achieved for C_{org} and N_{tot}, respectively (table 6.2). The calibrations based on the first derivative data required less PLS factors (three to four) compared to the calibrations based on absorbance data (six to seven factors). When looking at the estimates of the validation samples, the calibrations where wavelengths influenced strongly by water content were removed performed better compared to the calibrations including all wavelengths. Using all wavelengths resulted in clearly higher RMSE

6. Field measurements

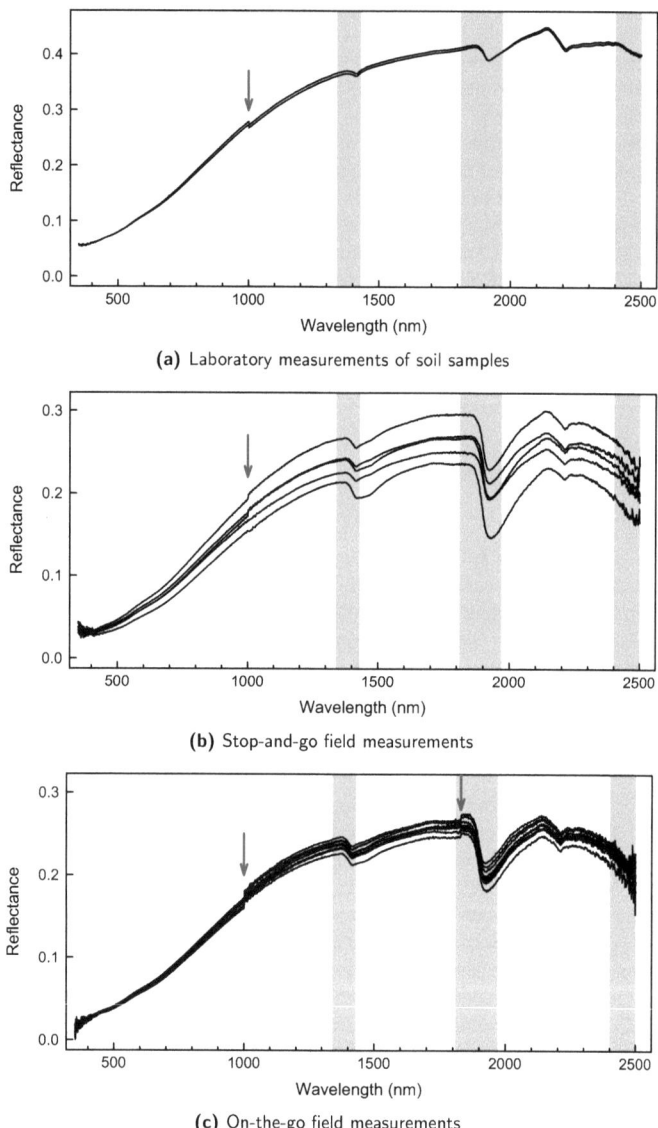

(a) Laboratory measurements of soil samples

(b) Stop-and-go field measurements

(c) On-the-go field measurements

Figure 6.6 Reflectance spectra collected at transect one from subplot two by different measuring setups. Blue coloured areas represent wavelengths influenced strongly by water content. Discontinuities due to sensor changes are designated with red arrows.

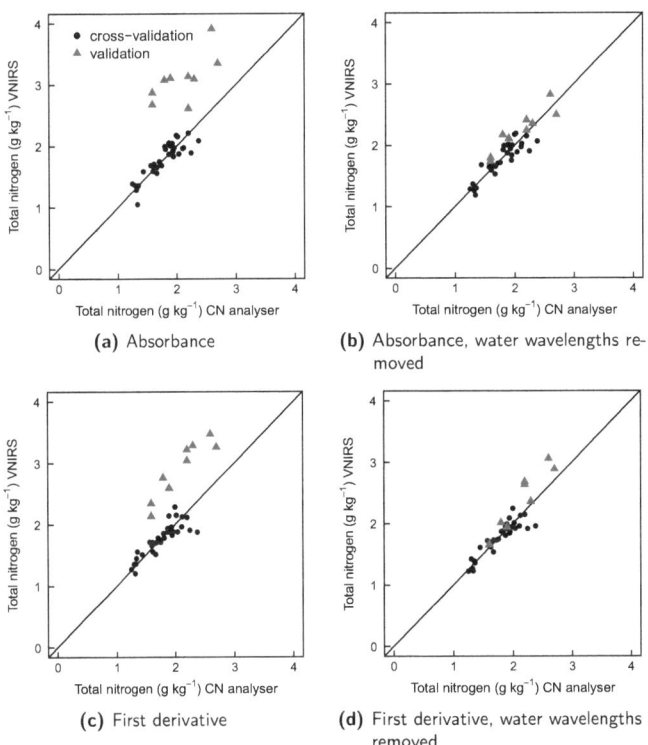

Figure 6.7 Cross-validation and validation N_{tot} estimates by VNIRS for calibrations based on the different spectral pre-treatments.

6. Field measurements

Table 6.2 Performance of VNIRS calibrations for Bad Lauchstädt depending on the spectral pre-treatments: number of used PLS factors and RMSE of six-fold cross-validation (CV; 33 samples) and validation (Val; nine samples)

Spectral pre-treatment	C_{org}			N_{tot}		
	Factors	$RMSE_{CV}$	$RMSE_{Val}$	Factors	$RMSE_{CV}$	$RMSE_{Val}$
		— (g kg^{-1}) —			— (g kg^{-1}) —	
A	7	1.2	4.5	6	0.14	1.04
A–w	7	1.3	3.6	7	0.13	0.21
dA	3	1.5	12.0	3	0.15	0.82
dA–w	4	1.4	7.0	4	0.13	0.29

A: absorbance, A–w: absorbance, wavelengths influenced strongly by water removed, dA/dA–w: first derivative of absorbance, with/without water wavelengths.

and bias (figure 6.7). Considering cross-validation and validation performance, the best results were achieved using absorbance data discarding wavelengths influenced by water and including seven PLS factors for the calibration (validation RMSE: 3.6 and 0.21 g kg^{-1} for C_{org} and N_{tot}, respectively). Even the samples with N_{tot} contents above the calibration range were estimated well by this calibration model. Thus, it was used for subsequent C_{org} and N_{tot} estimations for the laboratory and field VNIRS measurements from plot seven.

6.3.3. Laboratory measurements

In this and the following subsection, the C_{org} and N_{tot} estimates based on VNIRS measurements will be presented. The results for transect two will only be presented graphically if they deviated from the results for transect one.

The C_{org} and N_{tot} contents estimated from VNIRS measurements of dried and sieved soil samples collected from transect one were compared to the CN analyses by dry combustion of samples collected in autumn 2009 (figure 6.8; data provided by Helmholtz Centre for Environmental Research – UFZ Leipzig). Although the two sets of soil samples were collected at different times (autumn 2009 vs. April 2010) and from different soil layers (top 20 vs. top 5 cm), the VNIRS estimates showed the same spatial trends for C_{org} and N_{tot} than the CN analyser measurements. For both soil parameters, VNIRS produced higher values compared to dry combustion. The differences between the two methods were consistent over the whole transect. It will be discussed in section 6.4.4 whether the observed differences occurred due to differences in C_{org} and N_{tot} contents of the measured soil samples, or due to differences between the two analytical methods.

6.3.4. Field measurements

No differences were observed between the two measuring setups used in the field. Stop-and-go and on-the-go measurements both indicated similar spatial trends and variation for C_{org} and N_{tot} (figure 6.9). The field measurements showed clearly higher variation compared to the laboratory measurements. Additionally, the estimates based on field data were strongly biased to higher C_{org} and N_{tot} contents. The field estimates included a range from 40 to 80 g C kg^{-1} and from 3 to 6 g N kg^{-1}, whereas the laboratory estimates included a range from 20 to 30 g C kg^{-1} and

6.3. Results

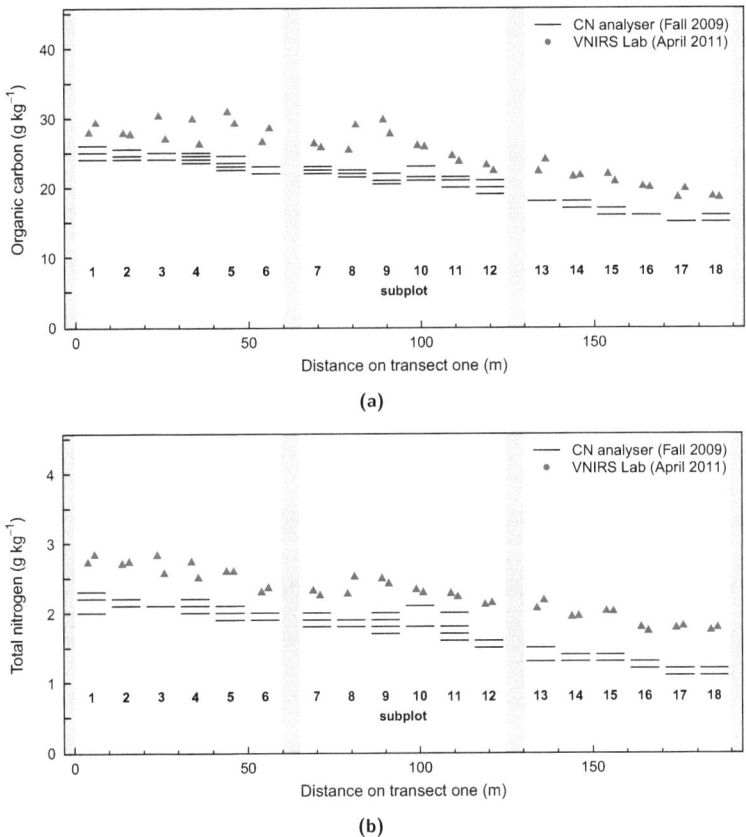

Figure 6.8 Comparison of (a) C_{org} and (b) N_{tot} contents of transect one, plot seven, determined by VNIRS laboratory measurements and CN analyser (four replicates per subplot; data provided by Helmholtz Centre for Environmental Research – UFZ Leipzig)

6. Field measurements

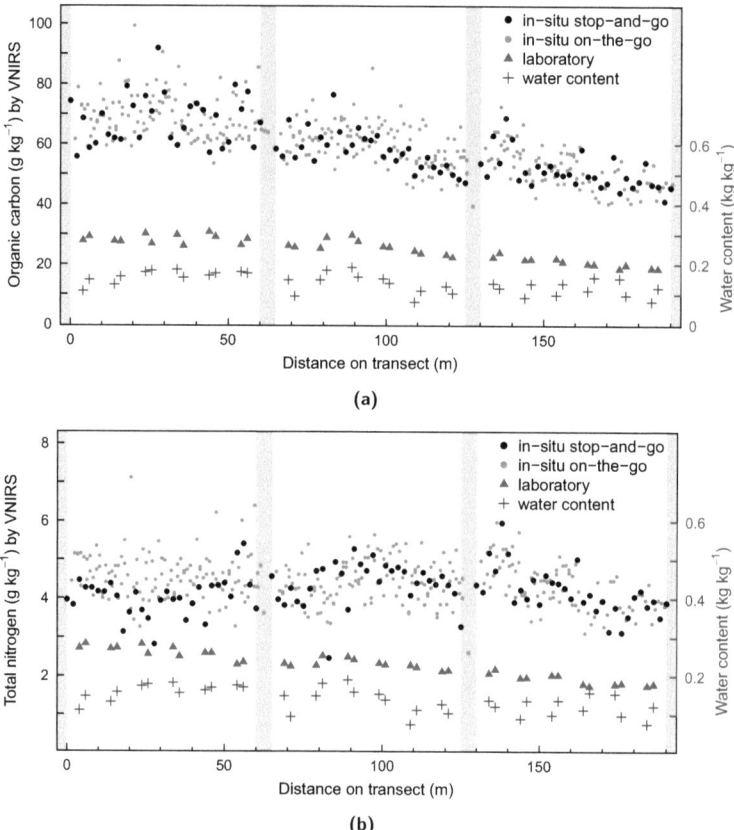

Figure 6.9 Comparison of (a) C_{org} and (b) N_{tot} VNIRS estimates from field (in situ) stop-and-go, on-the-go, and laboratory measurements of transect one; gravimetric water content determined for laboratory samples.

2 to $3\,\mathrm{g\,N\,kg^{-1}}$, respectively. While the field measurements reflected the spatial trend for C_{org} indicated by the laboratory measurements reasonably, there were clear discrepancies regarding N_{tot}, especially for the first half of the transect (0 to 90 m distance). For this part of the transect, N_{tot} in situ estimates seemed to correlate with the water content. Also for C_{org}, deviations between laboratory and field measurements were visible for the first third of the transect (0 to 60 m): while the laboratory measurements indicated only minor changes in C_{org} contents for this portion of the transect, field measurements indicated an increase (0 to 30 m distance on the transect) with a subsequent slight decrease (30 to 60 m). Again, there seemed to exist a correlation between the described deviations and water content.

The quality of C_{org} and N_{tot} estimates from field measurements was improved considerably by spiking the calibration with spectral data of nine samples collected from transect one (figure 6.10 a-d). The C_{org} and N_{tot} contents estimated by the spiked calibration models were comparable to the values determined in the laboratory. With the exception of few outliers, the spatial trends of C_{org} and N_{tot} over the transect were reflected well. By including both field and laboratory spectral measurements of the nine samples used for spiking, the variation of the C_{org} and N_{tot} estimates for the field measurements was clearly reduced compared to the calibration models that only included the field data of these samples. Also the number of outliers was smaller for the first.

The calibration models based only on the stop-and-go field measurements corresponding to the nine validation samples from transect one produced C_{org} and N_{tot} estimates comparable to the spiked calibration models (figure 6.10 e-f). They produced less outliers for the stop-and-go measurements than the spiked models, but the variation of the on-the-go measurements seemed to be slightly higher.

Surprisingly, all calibration models using field data – the spiked models as well as those using field spectra only – failed to estimate a part of the measurements taken from transect two. While the C_{org} and N_{tot} estimates between 100 and 190 m distance were estimated reasonably, the remaining were greatly underestimated. Transect two was measured in inverse direction, thus the measurements estimated well were collected first, and the problems occurred for the measurements collected at the end. For the calibration models spiked with field and laboratory spectra, the deviation from the reference measurements was smaller compared to the remaining.

6.4. Discussion

6.4.1. Visual screening

The differences observed between the spectra measured in the field and the laboratory were expected because the field measurements were taken from the field moist soil surface, whereas the laboratory measurements were taken from oven dried soil samples. As described in chapter 5 of this work, an increased water content has mainly two effects on soil reflectance: the reflectance is reduced over the whole range of visible and near infrared radiation, and the characteristic water bands near 1400 and 1900 nm wavelengths are broader and deeper. Consequently, it was concluded that the differences between field and laboratory spectra mainly were caused by differences in soil moisture, although other factors like soil surface roughness and the presence of plant residues and stones in the field might have accounted for some differences, too.

The field spectra exhibited more noise compared to the laboratory spectra. One part of the additional noise was assumed to be introduced by the field conditions. While the laboratory spectra were collected with a contact probe, the field measurements were illuminated and measured from

6. Field measurements

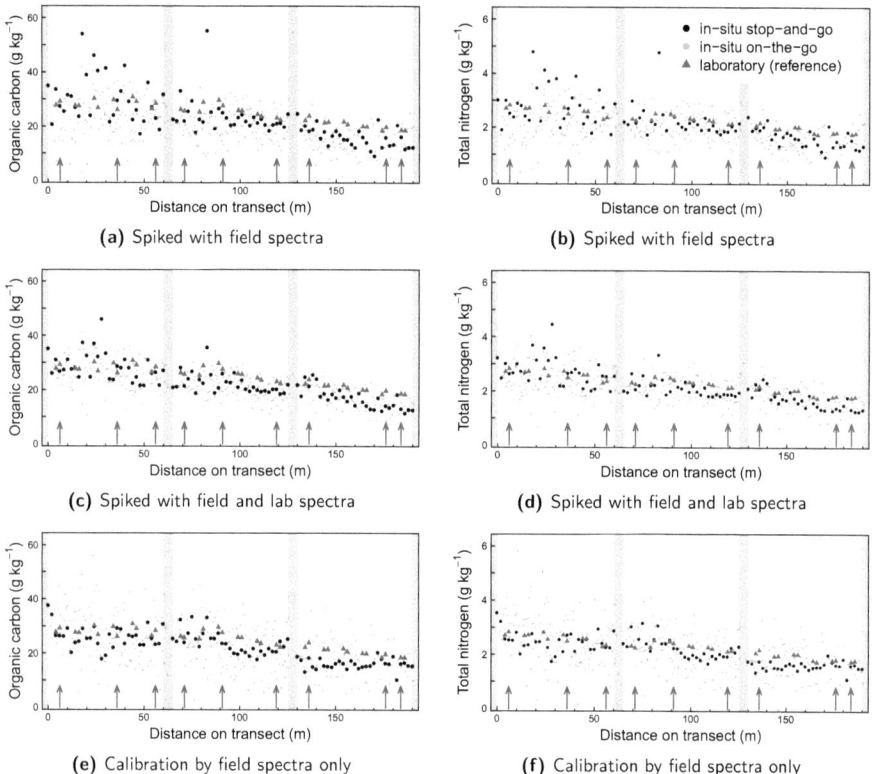

Figure 6.10 C_{org} and N_{tot} estimates for VNIRS field measurements of transect one using calibration models based on laboratory spectra spiked with field spectral data (a-b), spiked with laboratory and field spectral data (c-d), and using calibrations based only on field measurements of nine local validation samples (e-f). Red arrows indicate the locations on the transect where the samples used for spiking were taken. Laboratory VNIRS estimates (red triangles) were derived from calibration models without spiking (same values as in figure 6.9).

6.4. Discussion

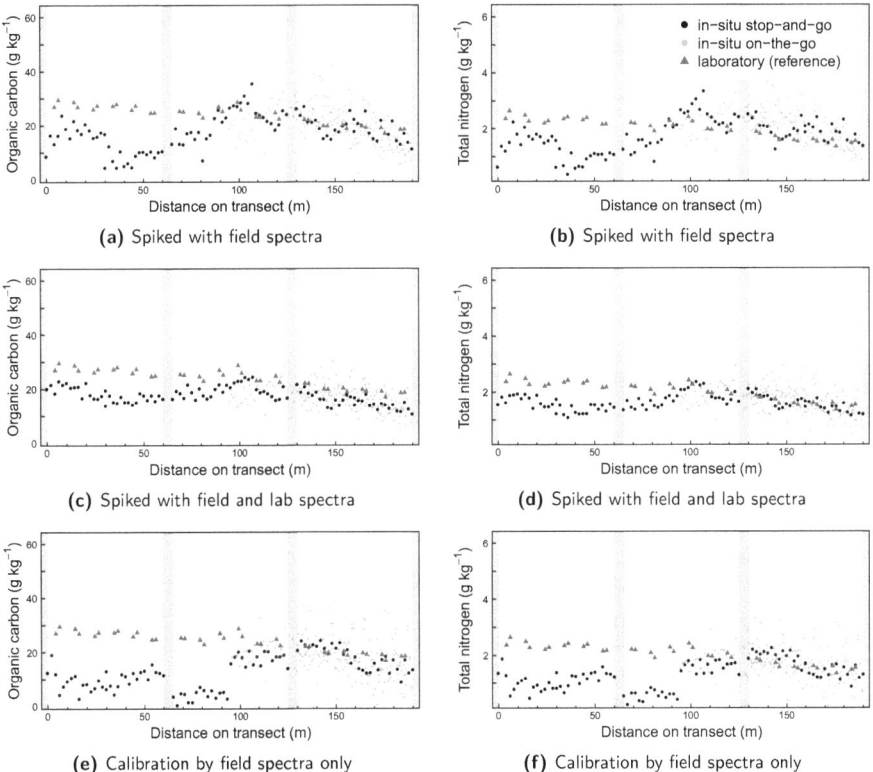

Figure 6.11 C_{org} and N_{tot} estimates for VNIRS field measurements of transect two using calibration models based on laboratory spectra spiked with field spectral data (a), spiked with laboratory and field spectral data (b), and using calibrations based on field measurements only of nine local validation samples (c-d). Red arrows indicate the locations on the transect where the samples used for spiking were taken. Laboratory VNIRS estimates (red triangles) were derived from calibration models without spiking.

6. Field measurements

a short distance. Generally, higher illumination intensities are achieved for contact measurements and they are influenced less by stray light, and therefore, the signal-to-noise ratio for the field measurements was expected to be worse. Furthermore, moving the platform during the on-the-go measurements caused vibrations and minor changes in the probe-to-soil distance and introduced additional noise. Also the fact that the steps at the sensor changes were promoted for the field measurements – especially for the on-the-go setup – was attributed to vibrations and distance changes due to the movement of the platform.

When comparing stop-and-go and on-the-go measurements, the first exhibited much more variation. The stop-and-go spectra did not differ in respect of their shape, but the whole spectra were shifted to higher or lower reflectance values. By comparison, the on-the-go spectra were much closer. During stop-and-go measurements, a rather small area of soil was screened, while during on-the-go measurements, a much bigger soil area was screened due to the movement of the platform. Thus, the on-the-go measurements averaged over a bigger area and smoothed variations – caused either by variability of the soil, or by errors introduced by the measuring setup. Considering the variations of C_{org} and N_{tot} contents observed for the four replicate soil samples per subplot (figure 6.5) indicated that, even at small scales, soil properties are varying strongly within the measured plot. Averaging over some minimum amount of soil (or area) is required to achieve representative and comparable results. Assuming that most of the variability of the stop-and-go was really due to soil variability, I concluded that on-the-go measurements were more appropriate for the mobile platform used for this project – except if the soil variation at small scales, say at intervals of 10 to 20 cm, was of special interest.

6.4.2. Calibration

The reported cross-validation and validation results confirmed that VNIRS calibrations can be established for local scale areas and used for soil samples collected later. Interestingly, only the calibration models for which wavelengths influenced strongly by water had been removed were able to estimate the validation samples well, whereas all tested models performed similarly in cross-validation. Consequently, it is necessary to validate an existing calibration by few reference measurements when applied to a new set of soil samples. The big validation bias for the calibration models using all wavelengths was possibly caused by minimal differences in soil moisture because there is always some interaction of the dried samples with air and the air moisture is varying greatly over time. Yet, an alternative explanation seemed reasonable too: the spectral properties of calibration and validation samples differed due to other factors than moisture – e.g. soil aggregation because of different soil treatment prior to sample collection – and coincidentally the differences mainly occurred in the range of the removed water wavelengths. The first explanation seemed more probable to me, and moreover, it was supported by the results discussed in section 5.4.1 (page 72). There, it has been shown that wavelength ranges influenced strongly by water exhibited reflectance changes even for very low moisture contents, whereas the remaining wavelength ranges exhibited stable reflectance values for low moisture contents. Therefore, for future calibrations, it should always be tested if they are more stable when water wavelengths have been removed. An important aspect is that real validation samples from a different ensemble of soil samples are needed instead of cross-validation to assess the effect of water wavelength removal correctly.

6.4. Discussion

6.4.3. Laboratory measurements

Comparing C_{org} and N_{tot} estimates based on VNIRS laboratory measurements with traditional CN analyses of soil samples collected several months earlier (figure 6.8) showed good agreement in respect of the spatial trends. The variability of the replicate samples per subplot was similar for both methods. VNIRS measurements resulted in higher C_{org} and N_{tot} contents for all samples. Either the samples analysed by VNIRS really exhibited higher contents, or there was a bias related to the analytical technique. The samples used for the traditional CN anayses were collected in autumn 2009, while the samples used with VNIRS were collected in April 2010. Leinweber et al. (1994) collected samples from plots of the long-term fertilisation experiment Bad Lauchstädt during a year and showed that C_{org} and N_{tot} contents as well as the C/N ratio varied strongly over time. The extent of the seasonal variations depended on the fertiliser input, but generally C_{org} and N_{tot} contents were reduced by 10 to 15 % in autumn compared to spring (Leinweber et al., 1994, figure 2). This was consistent with our measurements and explained most of the differences in C_{org} and N_{tot} contents between the samples collected in autumn and spring, although the differences seemed to be slightly bigger for our samples, especially for N_{tot} and subplots with low fertiliser input. Moreover, there was a second parameter contributing to the differences between the two ensembles of soil samples: the samples collected in autumn included the top 20 cm, whereas those collected in spring included only the top 5 cm of the soil. Even if the plough layer of the sampled plot had been mixed regularly due to tillage, some accumulation of organic matter might have occurred near the soil surface leading to increased contents of C_{org} and N_{tot} for the top 5 cm soil samples. Based on the precedent argumentation, it was concluded that the higher C_{org} and N_{tot} contents reported for the VNIRS measurements were attributable mainly to seasonal variations and, to some smaller extent, differences between the sampled soil layers – but they were not due to differences between the two analytical techniques. Thus, the results of the VNIRS laboratory measurements were considered accurate and appropriate to validate the VNIRS field measurements.

6.4.4. Field measurements

Applying the models calibrated by the dried soil samples to the field data resulted in C_{org} and N_{tot} estimates that correlated mostly with the reference measurements in respect of the spatial trends, but were shifted to much higher values. Apparently, these calibration models were not suitable for the field measurements because the spectra produced by the field and laboratory setups were too distinct. As discussed in section 6.4.1, the differences were mainly attributed to soil moisture. It was supposed that the bias was proportional to the soil moisture. The ratio as well as the difference between field and laboratory estimates were calculated and plotted against the gravimetric water content (see figure 6.12 for C_{org}; the results for N_{tot} were very similar and are therefore not shown). The scatter plots did not give any evidence for a close relation between the soil water content and the bias of the field measurement. Either there exists no relation between the two, or the water content determined for the samples was not correlated to the soil moisture relevant to the VNIRS measurements. The sensor measured the soil surface, and thus the moisture of the surface was relevant and might have differed from the moisture of the top 5 cm. Summing these findings up, it was concluded that calibration models based on spectral laboratory measurements cannot be used with field data, and there is no straightforward way to adapt them for field spectra without knowing the carbon contents of the samples, even if the soil moisture was known.

6. Field measurements

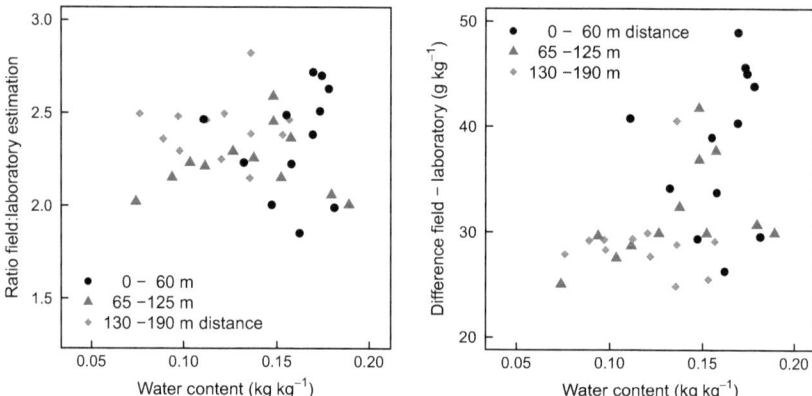

Figure 6.12 Ratio and difference of field and laboratory VNIRS estimates for C_{org} versus the gravimetric water content of corresponding soil samples of transect one.

Spiking the calibration models with spectral data of nine samples from transect one greatly improved the estimations. Both C_{org} and N_{tot} estimates of these models showed good agreement with the reference measurements. Although the field spectra accounted only for a small part of the calibration data, they sufficed to adapt the models to the field spectra. This was a confirmation of published results, e.g. by Viscarra Rossel et al. (2009) who also added few spectra collected in the field to a large collection of laboratory spectra (cf. section 6.1). Interestingly, adding both field and laboratory spectral data of the nine validation samples produced more accurate estimates and less outliers compared to the models spiked only with field spectra. By adding both field and laboratory spectra, the calibration model was able (if this can be said for a mathematical model) to compare corresponding field and laboratory data and derive a more exact model for the field data. Surprisingly, using only the nine field spectra for calibration (without any laboratory spectra) yielded results comparable to the spiked models. Obviously, a small number of reference measurements is sufficient for a homogeneous field. The higher variation of the estimates for on-the-go measurements was attributed to the fact that the models were derived from stop-and-go spectra and were more appropriate for this spectra than the on-the-go spectra. In summary, the main conclusion of these findings was: it is indispensable to use spectral field measurements to derive calibration models for field data, but only few field spectra are sufficient to spike an existing calibration based on laboratory spectra or to calibrate a model using field spectra only.

The problems observed with the estimates for transect two (figure 6.11) were either due to a problem with the field measurements for this part of the transect (0 to 100 m distance), or due to spectral features present in these spectra that were not covered by the calibration spectra. As no serious problems were recognisable when screening the spectra visually, the second alternative seemed probable. The problems were observed for the part of the transect that was measured last. The nine samples used for spiking and the calibration models based on field spectra all originated from transect one. It seemed that, for some time, the calibration models kept working for transect two, and then, suddenly, failed and were not appropriate anymore. Possibly, the spectra estimated badly exhibited different spectral features because the soil really differed, but this reasoning seemed not sound because the two transects showed good agreement for reference CN analyses and also the spectral laboratory data. Finally, time – or more precise: the succession of measurements – was assumed to be the key factor. During the field measurements, some pollution of the sensor probe, the lamp, and the white reference could not be prevented. While the pollution of the sensor probe and the lamp as well as the slight decrease in illumination intensity due to the discharge of the battery was corrected for by regular white reference measurements, the pollution of the white reference itself could not be corrected for. Possibly, the degree of pollution of the white reference excessed a certain level after measuring the first half of transect two leading to minor measuring errors with rather large effects on the C_{org} and N_{tot} estimates. It remained unclear why the effect was less promoted for the calibration models including both field and laboratory spectra than for the models using field data only. It seemed that the first were generally more stable. The lesson to learn from these results: if calibration models are spiked with field data or are derived from field data only, the used field spectra must cover on one hand the whole area of interest (spatially), but on the other hand also the whole time interval of measurements. While the first seems trivial and is also strongly recommended for laboratory data, the latter is specific for field measurements and the used measuring setup, but may also apply for other on-the-go sensors.

6.5. Summary and Conclusions

Within the present work, a portable FieldSpec 3 spectrometer was mounted to a mobile platform to conduct stop-and-go and on-the-go field measurements. A halogen lamp was used for illumination, and a cover excluded environmental light. Compared to laboratory measurements, field measurements exhibited a worse signal-to-noise ratio and spectral differences related to soil moisture like lower reflectance values and promoted water bands near 1400 and 1900 nm wavelengths. Calibration models based on 33 laboratory measurements of dried samples were not able to estimate C_{org} and N_{tot} contents for the field measurements correctly, but the spatial trends were reflected reasonably. Adding few field spectra to the calibration set *(spiking)* greatly improved the quality of the estimations. C_{org} and N_{tot} contents were estimated reasonably by all spiked calibrations, but adding both field and laboratory spectra of the nine samples used for spiking yielded better results than adding the field spectra only. Moreover, calibration models using only nine field spectra were derived. They yielded results comparable to the spiked models.

Based on the presented results, it was concluded:

- On-the-go measurements are integrating over a bigger soil area than stop-and-go measurements and should be preferred – except if the soil variability at small scales is of special interest.

- It should be tested for all calibration models if they are more stable when wavelengths related to water are removed. Real validation samples are needed for this purpose, as the effect is not necessarily visible in cross-validation.

- Due to the spectral differences related to soil moisture, calibration models based on laboratory spectra cannot be used directly to estimate field spectra.

6. Field measurements

- Only few field spectra of samples analysed by the reference method are needed to adapt an existing calibration to estimate the field spectra reasonably.
- Calibrations only using few field spectra yield results comparable to spiked models.
- The field spectra used for spiking and for calibrations should represent the area of the field measurements *spatially*, but also the whole *time period* of the field measuring campaign.

7. Conclusions and perspectives

The results presented and discussed in the previous chapters demonstrated the big potential of VNIRS for soil analyses in laboratory and field applications. This chapter provides the conclusions derived from the achieved results as well as perspectives on desired further developments in the field of soil spectroscopy.

7.1. Laboratory applications

From the variety of available calibration algorithms used in the field of spectroscopy, partial least squares regression (PLSR) and wavelet transforms in combination with quadratic regression models were selected for the present work to estimate organic carbon (C_{org}) and total nitrogen (N_{tot}) from soil reflectance spectra. Both selected algorithms have in common that they first compress the data and then conduct a linear regression on the compressed data.[1] The performance (measured as cross-validation RMSE) strongly depended on the set of soil samples used and on the fine-tuning of the models. For both soil parameters assessed, increasing RMSE with increasing variability of the included soil samples was observed. For local and regional sets of soil samples, RMSE comparable to those reported by published studies were achieved (2 to 3 g C kg^{-1} and 0.2 to 0.3 g N kg^{-1}, respectively). For sets of national extension, higher RMSE than expected were observed, presumably due to the relatively small numbers of soil samples with respect to the soil variability included. C_{org} estimates of validation samples taken from another ensemble of soil samples than those used for calibration exhibited good correlation with the reference analyses, but the VNIRS estimates were biased. The performance of the models strongly depended on the number of PLS factors or wavelet coefficients, respectively, included in the model. Interestingly, the optimum number of factors or coefficients seemed to differ for cross-validation and independent validation.

For the assessment and monitoring of soil quality, for soil mapping, as well as for diverse scientific problems, analyses of various soil parameters are required. These are usually time-consuming and expensive. The results of the present work as well as published studies (Viscarra Rossel et al., 2006b; Stenberg et al., 2010) confirmed that VNIRS is a reliable alternative to conduct soil analyses at reasonable costs. Depending on the specific needs, local to regional VNIRS calibrations seemed to be suitable. For calibrations of larger extensions, comprehensive sets of spectral data are required. Because decreasing estimation accuracies were observed for more diverse datasets, the fragmentation of large datasets into more homogeneous sub-groups to derive separate calibrations seemed inevitable to me in order to achieve acceptable accuracy. The availability of large and comprehensive spectral libraries would permit to derive calibrations adapted to the set of analysed soil samples by selecting similar calibration samples from the collection. Currently, the Soil Spectroscopy Group is working on a global soil spectral library (Viscarra Rossel, 2009).[2] Another

---

[1] Mathematically, a quadratic regression model is considered as linear model because the quadratic term can easily be linearised by substituting the variable x^2 by $z = x^2$.

[2] http://www.proximalsoilsensing.org/global-spectroscopy/

strategy to adapt calibrations to a homogeneous set of soil samples was demonstrated in this work (section 3.3): the bias of the estimates is estimated by few reference analyses and all estimates are corrected for the bias.

Although good and promising results were achieved by diverse groups and scientists, VNIRS soil analyses are still far from being accepted by the majority as equivalent alternative for traditional analytical techniques. In my opinion, this is mainly due to the absence of standard procedures accepted by all soil spectroscopists as well as the fact that very often calibrations are not easily transferable from one spectrometer to another. These topics will be further discussed in section 7.3. Moreover, further improvements with respect to the stability of VNIRS calibrations are desirable. Corresponding suggestions derived from this work will be discussed in section 7.4.

2. Field applications

The results of VNIRS field measurements by a FS3 device mounted to a mobile platform illustrated the potential, but also the related problems of proximal soil sensing and on-the-go sensors. Technically, field measurements implied no serious problems, the real challenge was their interpretation. Spectra collected from field moist soils differed strongly from those of dried samples, mainly due to absorption by water molecules. While relative differences of C_{org} and N_{tot} were visible from on-the-go field measurements even when using a calibration derived from dried samples, specific calibrations were required to retrieve absolute levels. Furthermore, it was demonstrated that, for a homogeneous field, few samples (e.g. as few as nine) were sufficient to derive a calibration or adapt an existing calibration.

The results presented by this work confirmed that VNIRS on-the-go sensors could become a valuable and useful tool. At this time, such devices are already commercially available (e.g. from Veris Technologies). In my (humble?) opinion, precision farming exhibits the biggest potential to take profit of on-the-go sensors. To adapt inputs of fertilisers and other resources, the cultivated fields must be captured comprehensively. Besides, the acquisition of the spectral data can be integrated in other processes, e.g. drilling or manuring. For some purposes, even the interpretation of the spectral data can be automated (e.g. see Maleki et al., 2008). Moreover, regularly collected spectral data can be used to monitor the evolution of the cultivated soils over the years. The applicability of VNIRS for farming could be further promoted by providing devices with lower spectral resolutions and/or constricted wavelength ranges at lower costs. The performance of such sensors would be sufficient by far for most precision farming applications.

While on-the-go sensors seem to be promising for small scale applications like precision farming, they are considered less adequate for large scale applications like soil mapping. Of course, comprehensively measured fields could be used for soil mapping, but if larger regions should be covered, there are much more efficient approaches to capture them reasonably by VNIRS. In section 7.5, such an approach will be presented.

7.3. The need for standardisation

During the last two decades, numerous studies were conducted assessing VNIRS applications. Numerous procedures were used to prepare the soil samples prior to the spectral measurements, to collect spectral measurements, to analyse the retrieved spectral data, and to judge the accuracy of the analyses. To overcome the experimental stage and become an analytical technique equivalent to the traditional ones, the soil spectroscopy community must agree on standards. The – to my knowledge – first study targeting this issue was published only recently by Pimstein et al. (2011). The authors of the study demonstrated that differing measuring procedures introduced considerable spectral deviations which where greatly reduced when applying a standardised measuring procedure. Therefore, the measuring procedure must be standardised, otherwise the benefits of comprehensive spectral libraries will be unnecessarily degraded.

A further deficit of soil spectroscopy is the absence of universally accepted internal standards. The results of the present work revealed that differences between spectra collected by different spectrometers occur which cannot be corrected for by usual white reference measurements using a Spectralon panel. Besides, these panels are prone to pollution. Therefore, the implementation of additional internal standards is strongly recommended. Pimstein et al. (2011) assessed different materials as possible internal standards and recommended bleached sand (carbonates and iron oxides were removed and the sand subsequently washed) as internal standard. Given that these results will be confirmed by further studies, the availability of this internal standard would be an important improvement. Furthermore, it is suggested that the commonly applied calibration using one single Spectralon panel should be revisited. For most analytical methods requiring calibration, a series of calibration samples covering the whole range of the target variable is used. It seems obvious that also VNIR spectrometers would retrieve more exact spectral data when calibrated with a series of reflectance targets (e.g. 10, 25, 50, 75, 90, and 100 % reflectance) instead of only one target representing 100 % reflectance.

Standardisation of the measuring procedure and the implementation of internal standards and more reliable spectral calibration would enhance the comparability of spectral data collected by different VNIRS devices, but also of spectral data collected over time by the same device. The comparability of spectral data is an essential prerequisite to establish comprehensive spectral libraries and to establish calibrations that are stable on the long term. Besides, no one should forget that spectrometers or parts of them may break down. In order to avoid that years of labour may become worthless within one moment, the suggested standardisation is needed.

7.4. Model stability

As previously discussed, the stability over time and between different spectrometers is crucial to guarantee reliable and accurate estimates of soil properties by VNIRS. In the previous section it was outlined that standardisation of spectral measurements was expected to improve the stability of VNIRS calibrations. The results of the present work further suggested that adaptations of the calibration procedure could also make them more stable and reliable. On one hand, discarding some problematic wavelengths from calibrations was suggested. On the other hand, it was advised to consider also model stability and not only model accuracy for VNIRS calibrations.

In chapter 5, it was shown that reflectance of some wavelengths started to decrease (compared to oven dry condition) for very low soil water contents already, while it remained stable for other wavelengths as long as the water content did not exceed a certain level. The wavelengths located inside the absorption features of water near 1400 and 1940 nm as well as wavelengths above 2400 nm were identified as those most sensitive to water content changes. It was suggested that removing these wavelengths could make calibrations more stable because the influence of minor changes in soil moisture (e.g. due to changing air humidity) would be reduced. While removing of wavelengths related to water was reported in various cases for analyses of spectra collected from moist soil samples, it was – to my knowledge – not suggested for spectral analyses of dried soil samples so far. The implementation of the suggested wavelength removal is hindered by the fact that its benefit will very often not be visible in cross-validation. Generally, spectral deviations introduced by minor differences in soil moisture tend to be bigger for independent validation samples than within the calibration set. The assumed improvement of the resulting models were expected to become more important on the long term – as demonstrated by the calibration presented in chapter 6 where significant improvements were achieved for validation samples collected one and a half year after the calibration samples. Because its mode of action is easily comprehensible regarding the results of the present work, the removal of water sensitive wavelengths (1340 - 1430, 1810 - 1970, 2400 - 2500 nm) was recommended if the remaining wavelengths provide enough information to estimate the targeted soil properties without or only minor losses in accuracy. Otherwise, it should be tested if at least a part of the mentioned wavelengths could be skipped.

Very often, different calibration algorithms are compared to select the best. Furthermore, most algorithms need fine-tuning, e.g. selecting the number of PLS factors. Regarding the presented results it was suggested that, in addition to estimation accuracy, the stability of the resulting models should be considered in order to improve their performance on the long term and to reduce the influence of random errors in the spectra. It was proposed that the reproducibility of replicate analyses – quantified by the standard error of laboratory (SEL) – was assessed. Thus, good models should, in addition to low RMSE, also exhibit low SEL. Moreover, replicate spectra collected from separate sub-samples of the soil samples could be included in the calibration set to make calibrations less sensitive to random errors. If desired, replicate analyses can be conducted by different spectrometers to amplify the applicability of the correspondent calibration. Additionally, future investigations should assess the benefit of robust regression methods for soil spectroscopy (e.g. robust methods for PLSR as presented by Hubert & Vanden Branden, 2003).

7.5. Soil mapping

The present work was conducted in the framework of the iSOIL project which focused on 'improving fast and reliable mapping of soil properties, soil functions and soil degradation threats' (iSOIL, 2008). VNIRS on-the-go sensors were considered useful for small-scale applications like precision farming, but of limited use for larger scales as required for soil mapping. To capture larger areas, retrieving information from a large number of locations scattered about the area seems more efficient than collecting high-resolution data of single plots. Besides, soil parameters like C_{org} are more accurately estimated from dried soil samples.

An approach implementing laboratory spectroscopy in soil mapping is presented in figure 7.1. First, samples are collected from the whole area, dried, sieved, and measured by VNIRS. The sampling locations may be selected by a systematic random scheme, or a stratified random scheme if auxiliary data (for example elevation model, land use, geological maps, ...) are available. Subsequently, the spectral data are used to assess the soil variability within the targeted area and to identify sub-groups within the soil samples. Based on this knowledge, it is decided if additional samples are needed for the whole area or parts of it. If appropriate (depending on the type of required maps), the samples for reference analyses are selected based on the spectral data. The selected samples should cover evenly the observed spectral diversity, a possible selection procedure, Latin Hypercube Sampling, was described by Viscarra Rossel et al. (2008). Based on the reference analyses, these soil parameters can be estimated for the remaining soil samples. If

7.5. Soil mapping

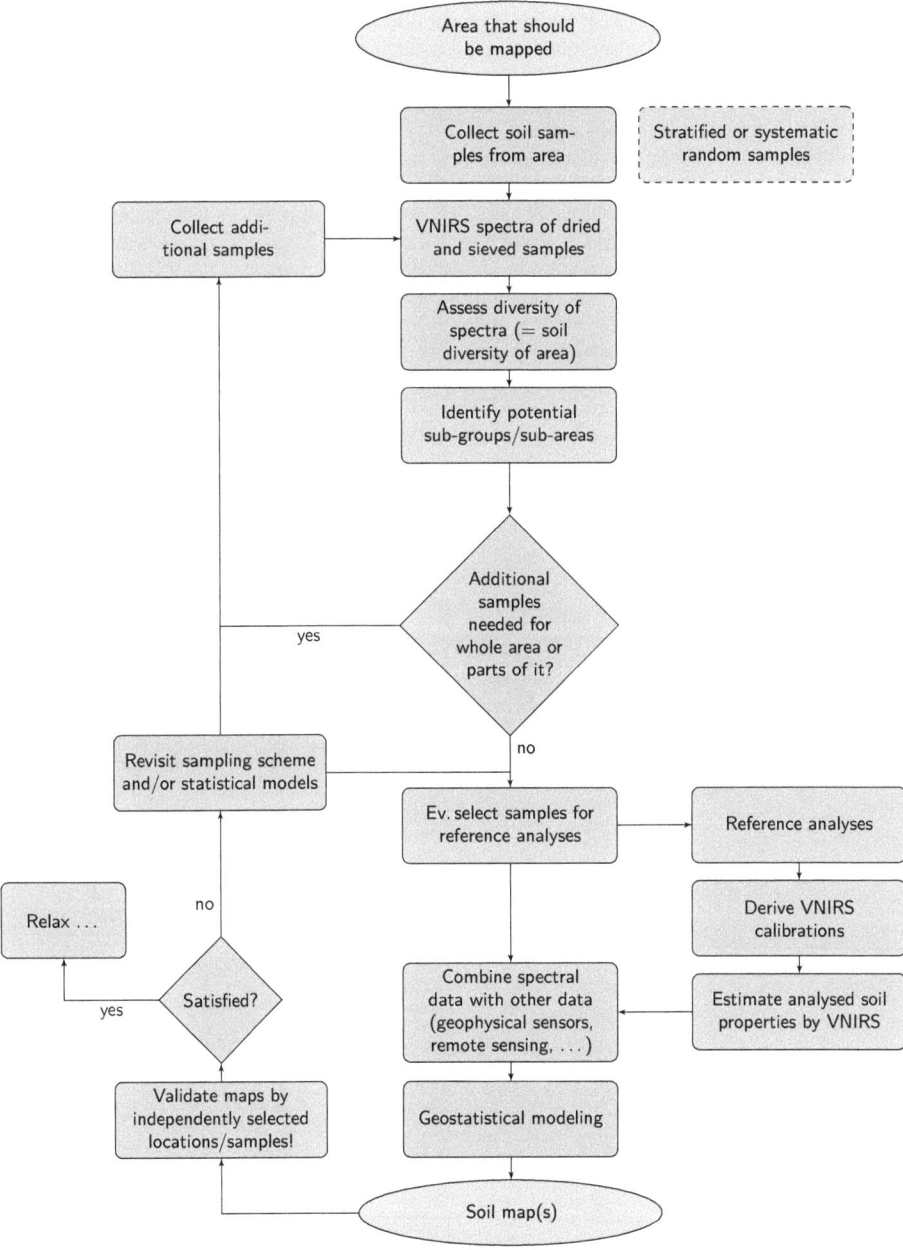

Figure 7.1 Flow chart: Proposed procedure to implement VNIRS in soil mapping.

7. Conclusions and perspectives

clearly differing sub-groups are present, the establishment of separate VNIRS calibrations for them should be considered. The spectral data and soil parameters derived from them are then combined with data from geophysical sensors, remote sensing, etc. Finally, soil maps are generated from the data by geostatistical methods. Of course, the resulting maps must be validated by independently selected locations and/or soil samples.

The presented scheme includes, on one hand, a holistic approach by assessing soil variability directly from the spectral data. On the other hand, soil parameters are estimated from the spectral data and used as well to produce soil maps. The described procedure is just one among a variety of thinkable implementations of VNIRS in soil mapping. The potential users are encouraged to use all of their fantasy to adapt it to their specific needs. For example, it could be – in some cases – more practicable to collect the spectral data directly in the field when visiting the sampling locations instead of carrying samples to the laboratory. Because of their relatively low costs, VNIRS analyses – in combination with other data – could become a valuable tool for soil mapping enabling the generation of comprehensive and reliable soil maps.

A. Additional information

A.1. Chapter 2: Methods

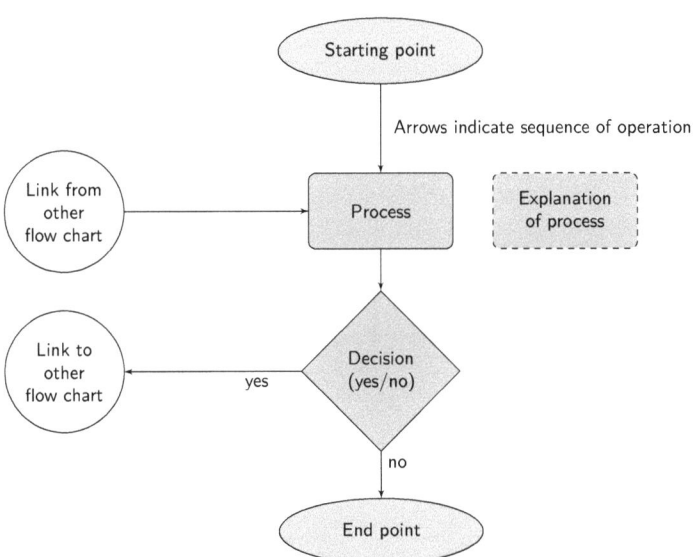

Figure A.1 Elements used for flow charts.

A. Additional information

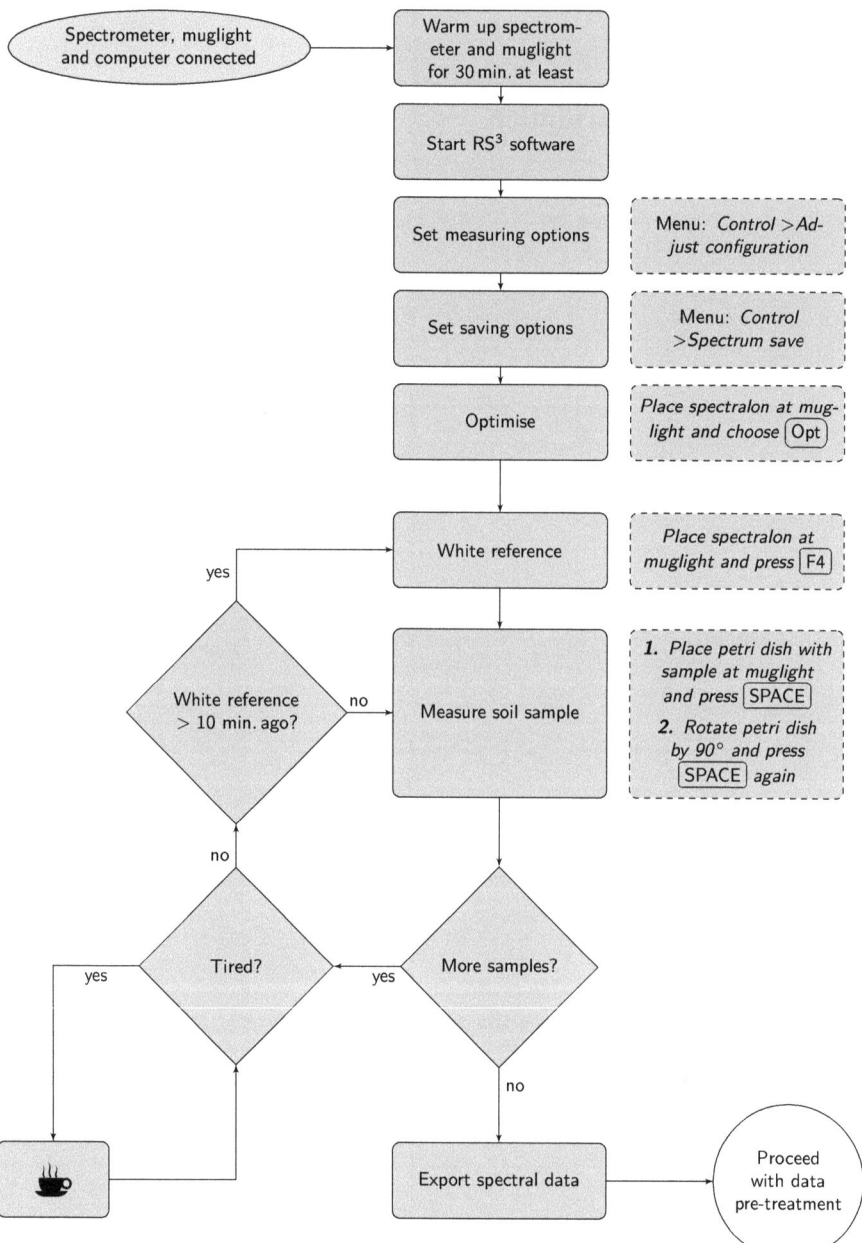

Figure A.2 Flow chart: Procedure for laboratory VNIRS measurements.

A.1. Chapter 2: Methods

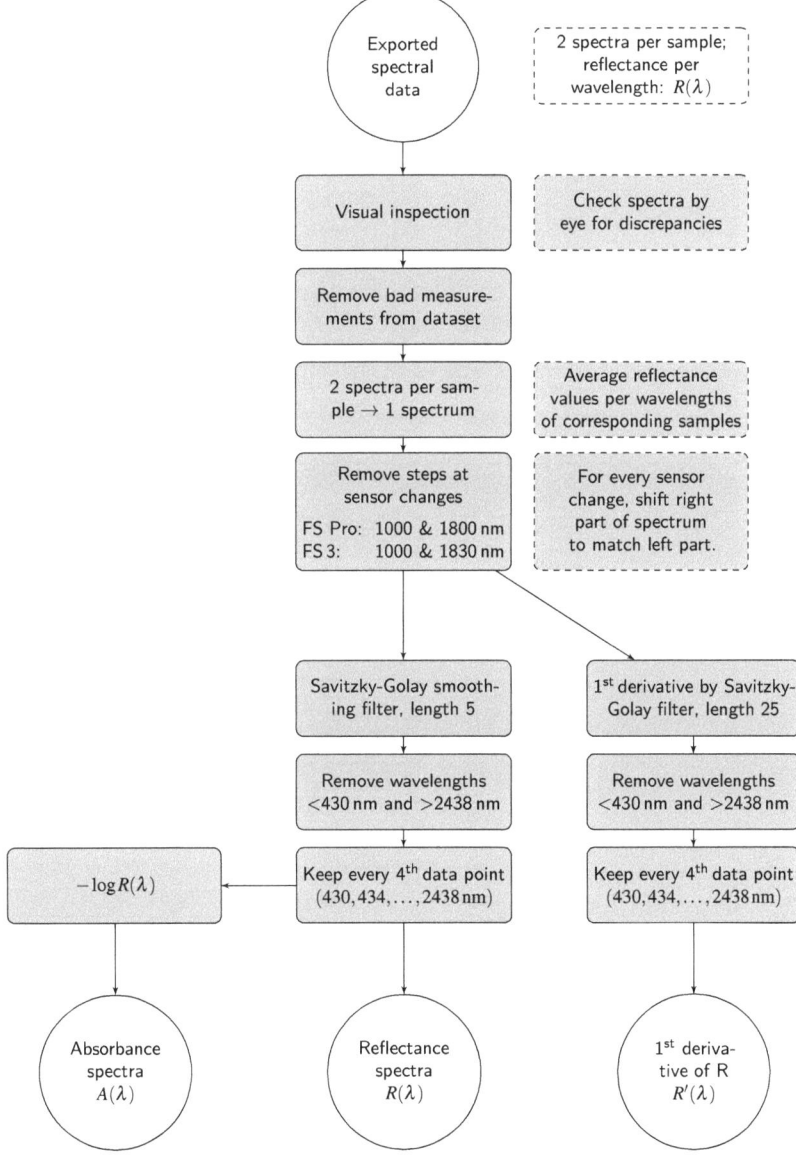

Figure A.3 Flow chart: Pre-treatments of spectral data.

A. Additional information

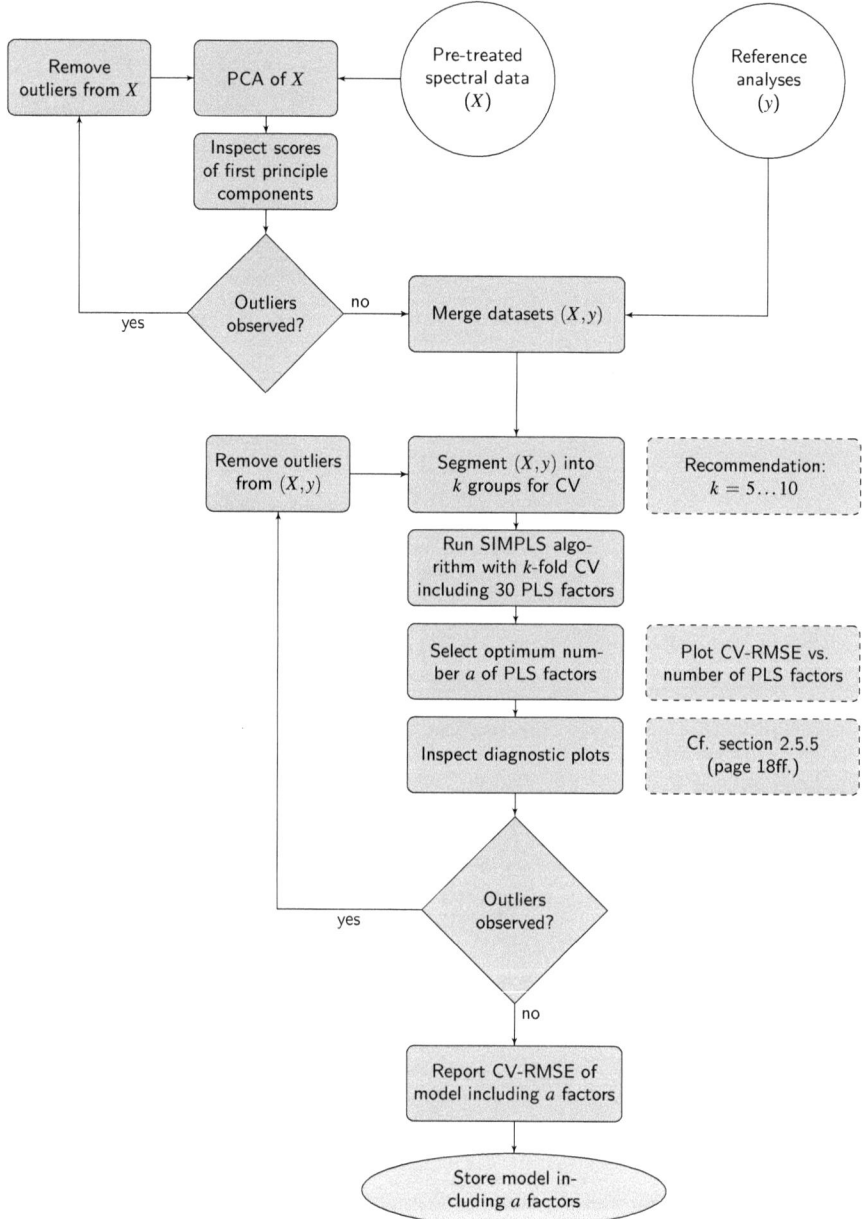

Figure A.4 Flow chart: Calibration by partial least squares regression (PLSR) for target variable y. CV: cross-validation; PCA: principle component analysis; RMSE: root mean squared error; SIMPLS: PLSR algorithm, cf. section A.1.1.

A.1. Chapter 2: Methods

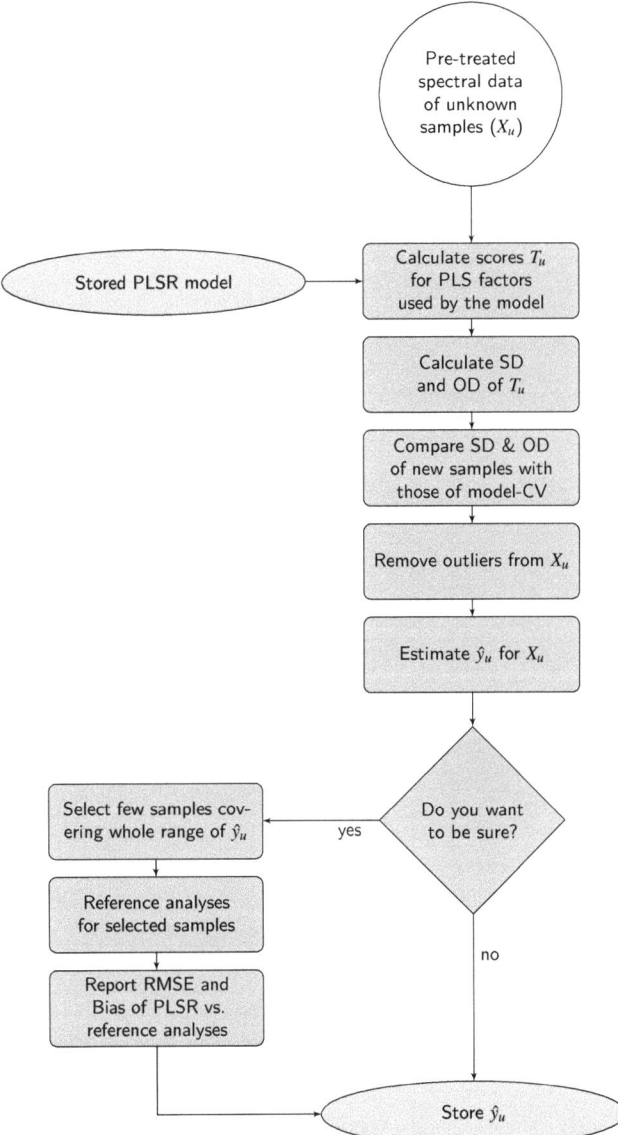

Figure A.5 Flow chart: Estimation of unknown samples by PLSR. CV: cross-validation; OD: orthogonal distance; RMSE: root mean squared error; SD: score distance.

A. Additional information

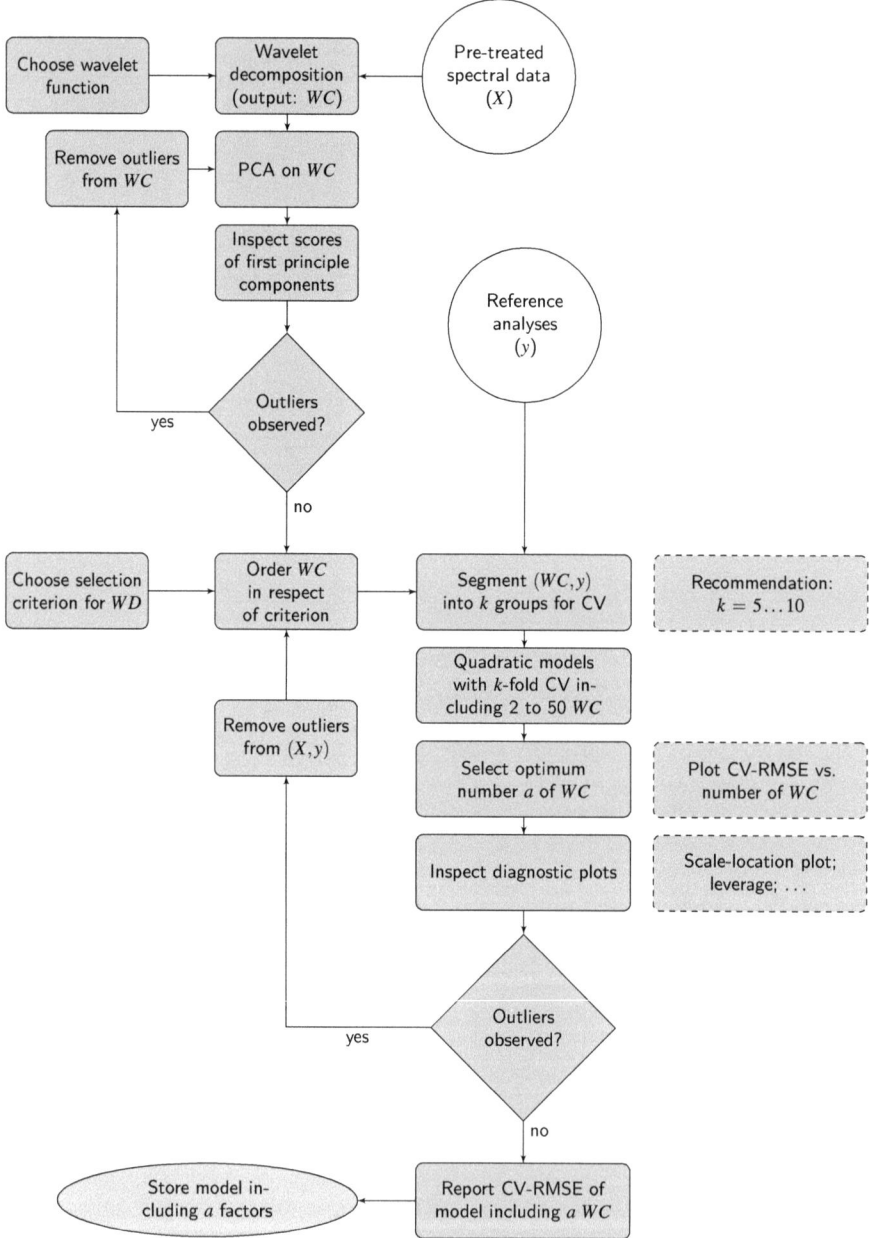

Figure A.6 Flow chart: Calibration for target variable y by models based on wavelet coefficients. CV: cross-validation; PCA: principle component analysis; RMSE: root mean squared error; WC: wavelet coefficients.

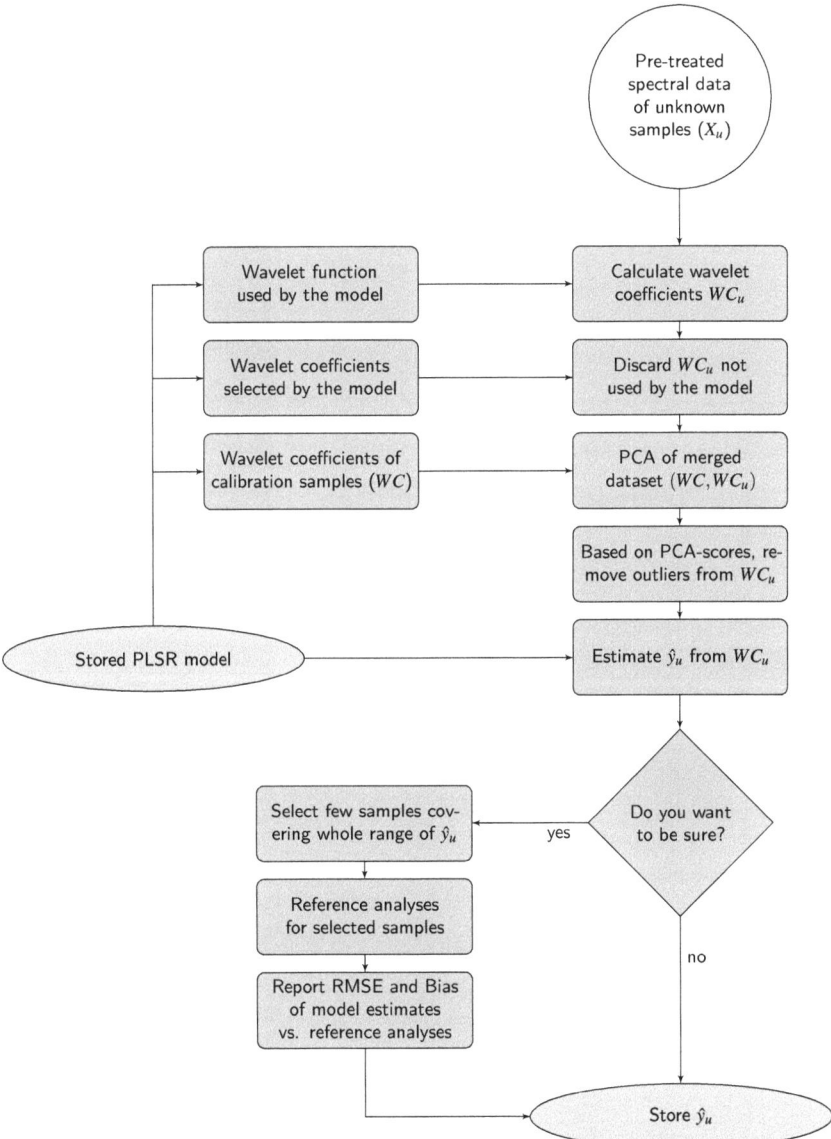

Figure A.7 Flow chart: Estimation of unknown samples by a model based on wavelet coefficients. CV: cross-validation; PCA: principle component analysis; RMSE: root mean squared error; WC: wavelet coefficients.

A. Additional information

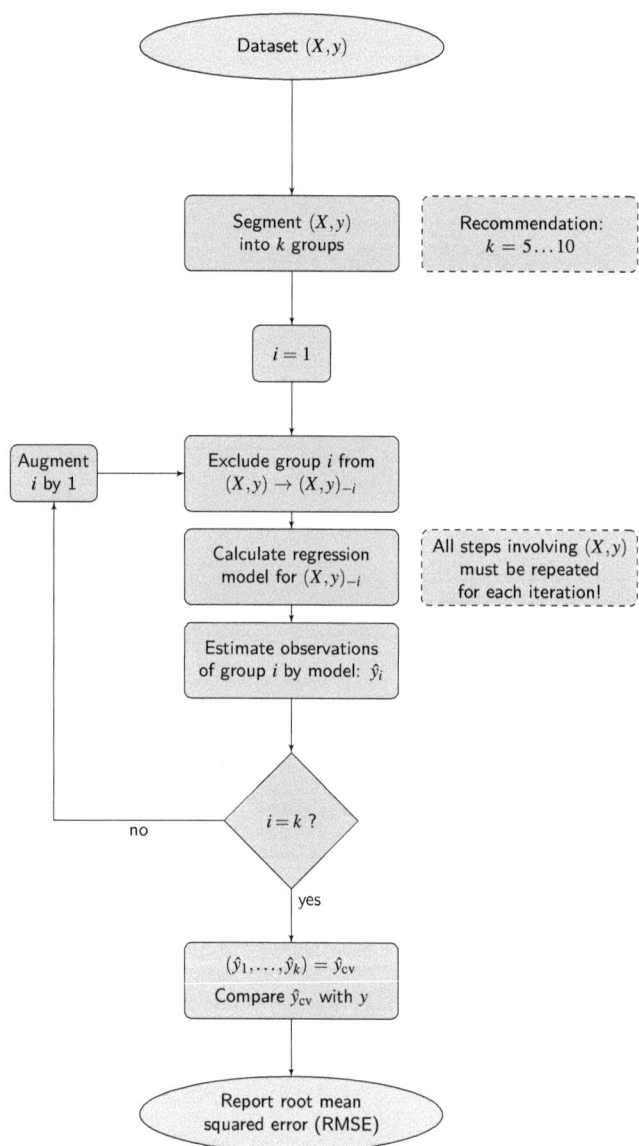

Figure A.8 Flow chart: Principle of cross-validation for regression models.

A.1. Chapter 2: Methods

A.1.1. SIMPLS algorithm

Pseudo-code for the SIMPLS algorithm for partial least squares regression (reproduced and modified from de Jong, 1993). The asterisk ($*$) designates matrix multiplication, and (^{T}) designates matrix transposition.

Input

$n \times p$ matrix X	n observations of p explaining variables, e.g. spectral data
$n \times m$ matrix Y	m independet variables, e.g. soil properties to estimate
Number of factors a	Number of dimensions to retain, $a \leq p$

Pseudocode

$Y_0 = Y - \bar{y}$	Center Y; $\bar{y}$ represents (column-wise) mean of Y
$S = X^T * Y_0$	Cross-product
For $i = 1,\ldots,a$:	
$\quad q =$ dominant eigenvector of $S^T * S$	Y block factor weights
$\quad w = S * q$	X block factor weights
$\quad t = X * w$	X block factor scores
$\quad t = t - \bar{t}$	Center t
$\quad normt = \sqrt{t^T * t}$	Compute norm
$\quad t = t/normt$	Normalise scores
$\quad w = w/normt$	Adapt weights accordingly
$\quad p = X^T * t$	X block factor loadings
$\quad q = Y_0^T * t$	Y block factor loadings
$\quad u = Y_0 * q$	Y block factor scores
$\quad v = p$	Initialise orthogonal loadings
$\quad$ *If $i > 1$ then*	
$\quad\quad v = v - V * (V^T * p)$	Make $v \perp$ previous loadings
$\quad\quad u = u - T * (T^T * u)$	Make $u \perp$ previous t^T values
$\quad$ *End*	
$\quad v = v/\sqrt{v^T * v}$	Normalise orthogonal loadings
$\quad S = S - v * (v^T * S)$	Deflate S with respect to current loadings ('subtract' current PLS factor)
$\quad$ Store $w, t, p, q, ,u,$ and v	
$\quad$ into W, T, P, Q, ,U, and V, respectively	
End	

Output

$B = W * Q^T$	Regression coefficients
$h = \text{diag}(T * T^T) + 1/n$	Leverage of objects
$varX = \text{diag}(P^T * P)/(n-1)$	Variance explained for X variables
$varY = \text{diag}(Q^T * Q)/(n-1)$	Variance explained for Y variables

Estimation of observation x_u

$\hat{y}_u = \bar{y} + (x_u - \bar{x})B$	$\bar{x}$ represents (column-wise) mean of X

A. Additional information

A.2. Chapter 3: VNIRS laboratory models

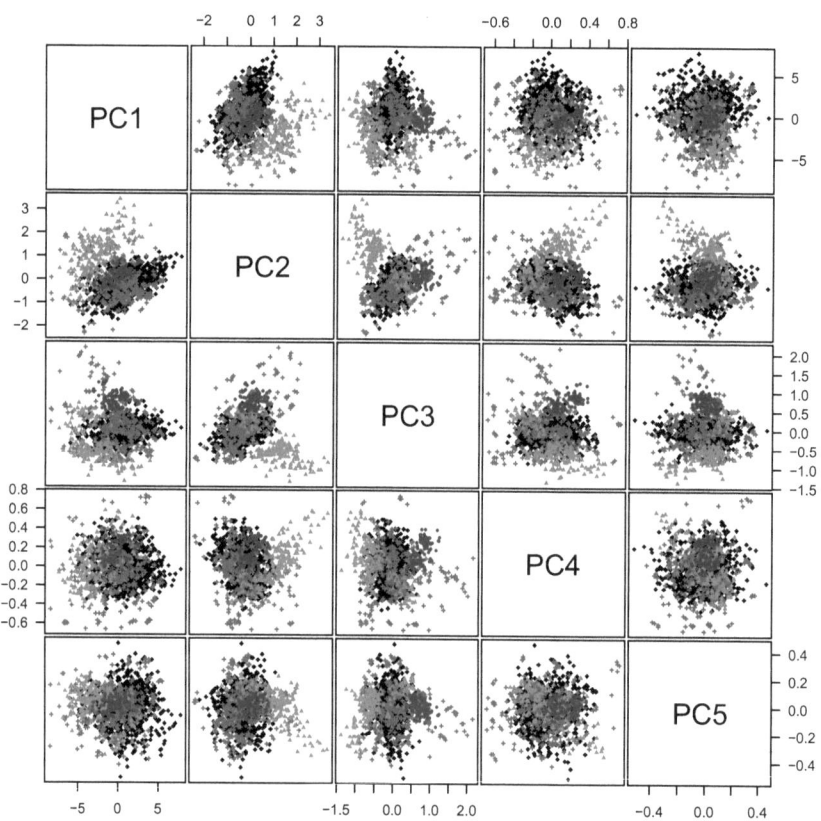

Figure A.9 Scatter plots of the first five principle components for the pooled dataset of all absorbance spectra used for the different PLSR models to estimate total and organic carbon: + *nabo-eic*, ◆ *fr-so*, ▲ *rosslau*, ● *lu*.

A.2. Chapter 3: VNIRS laboratory models

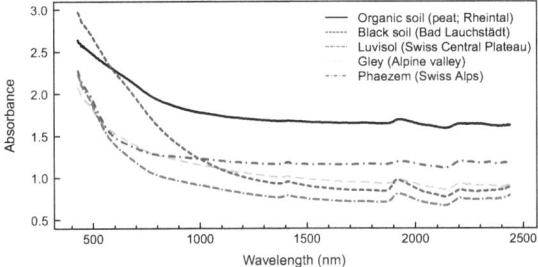

Figure A.10 Absorbance spectra of five randomly selected soil samples in oven dry condition (four samples from the *nabo* set and one from the *lauch* set; same samples as displayed in figure 2.2).

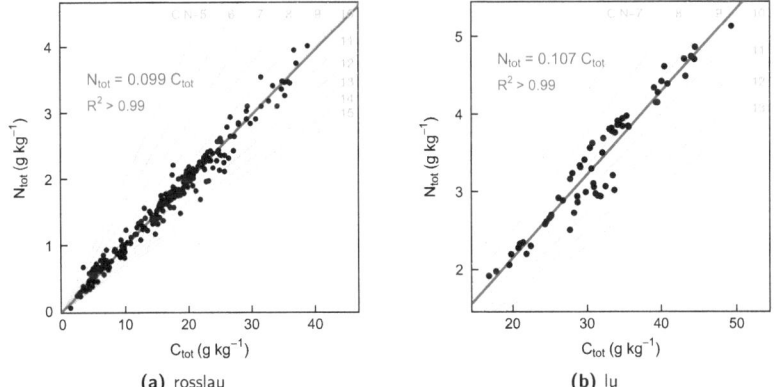

Figure A.11 Correlation of C_{org} and N_{tot} for the *rosslau* and *lu* datasets (cf. section 3.1.1).

A. Additional information

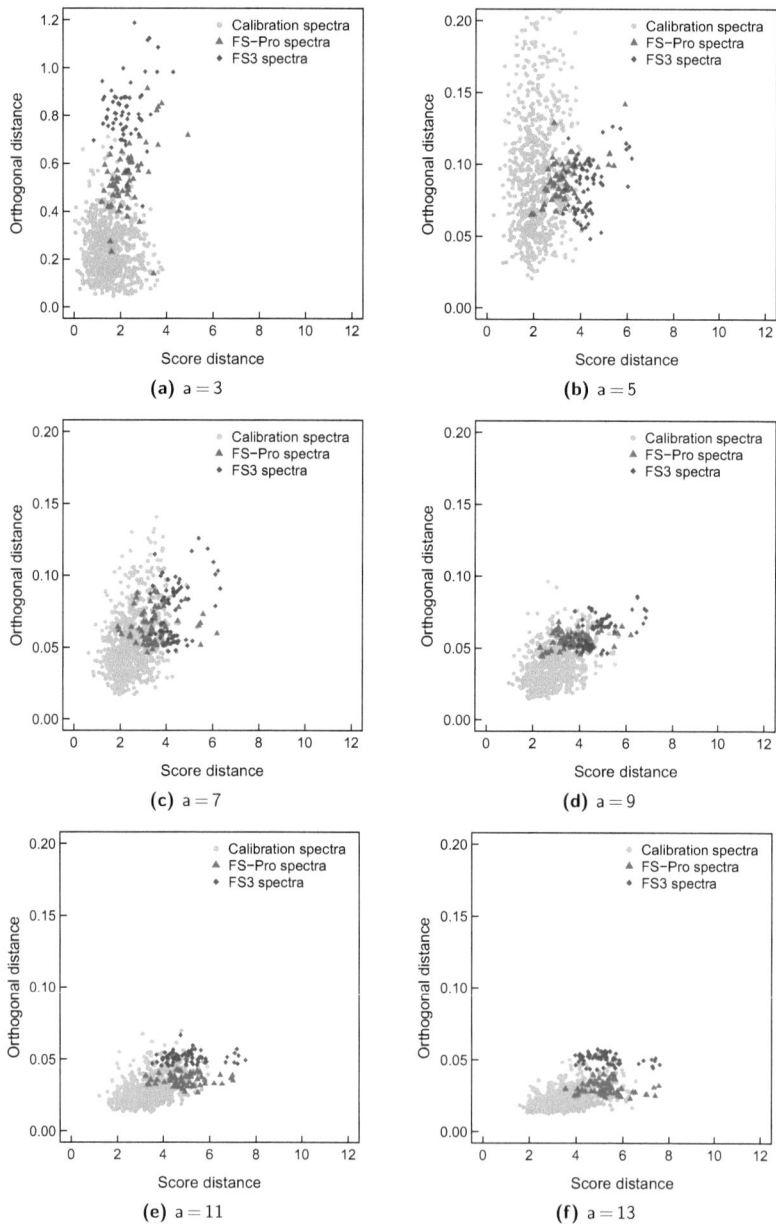

Figure A.12 Score distances and orthogonal distances of *lu* samples in respect of C_{org} models derived from the *fr-so* dataset including a = 3,5,...,13 PLS factors.

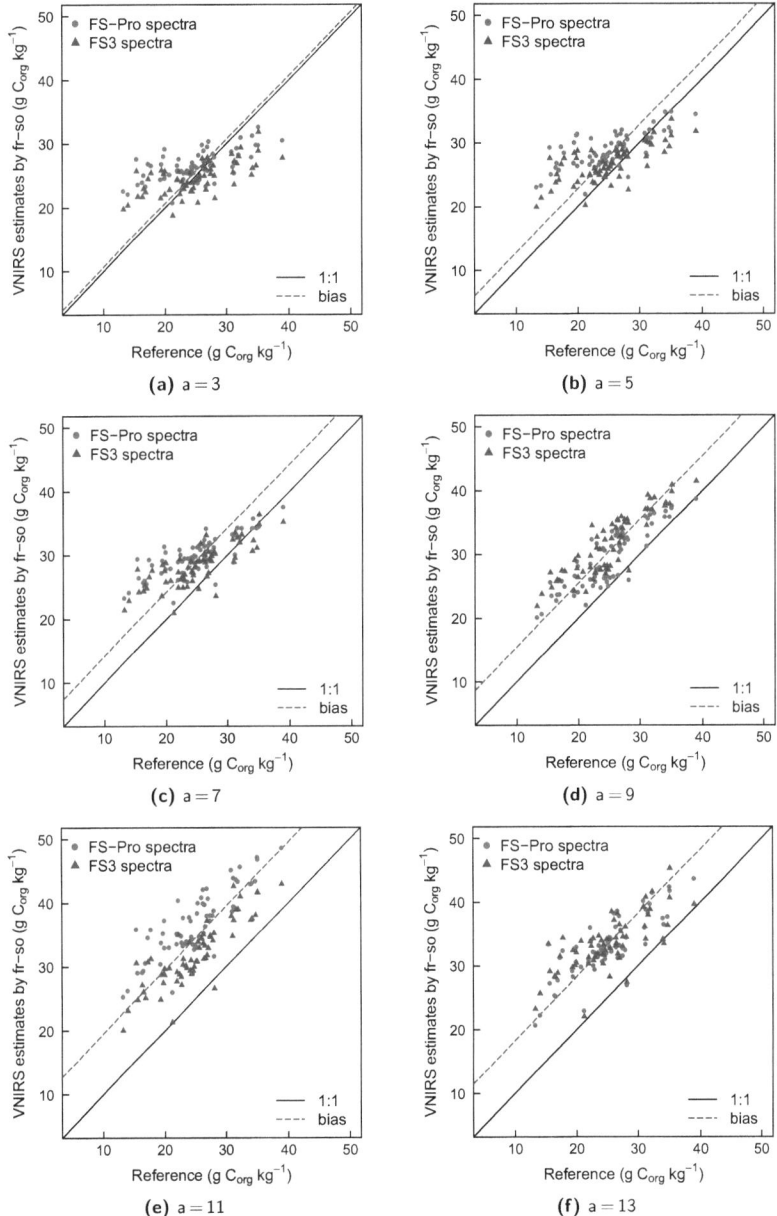

Figure A.13 Scatter plots of C_{org} estimates for *lu* samples by models derived from the *fr-so* dataset including a = 3,5,...,13 PLS factors.

A. Additional information

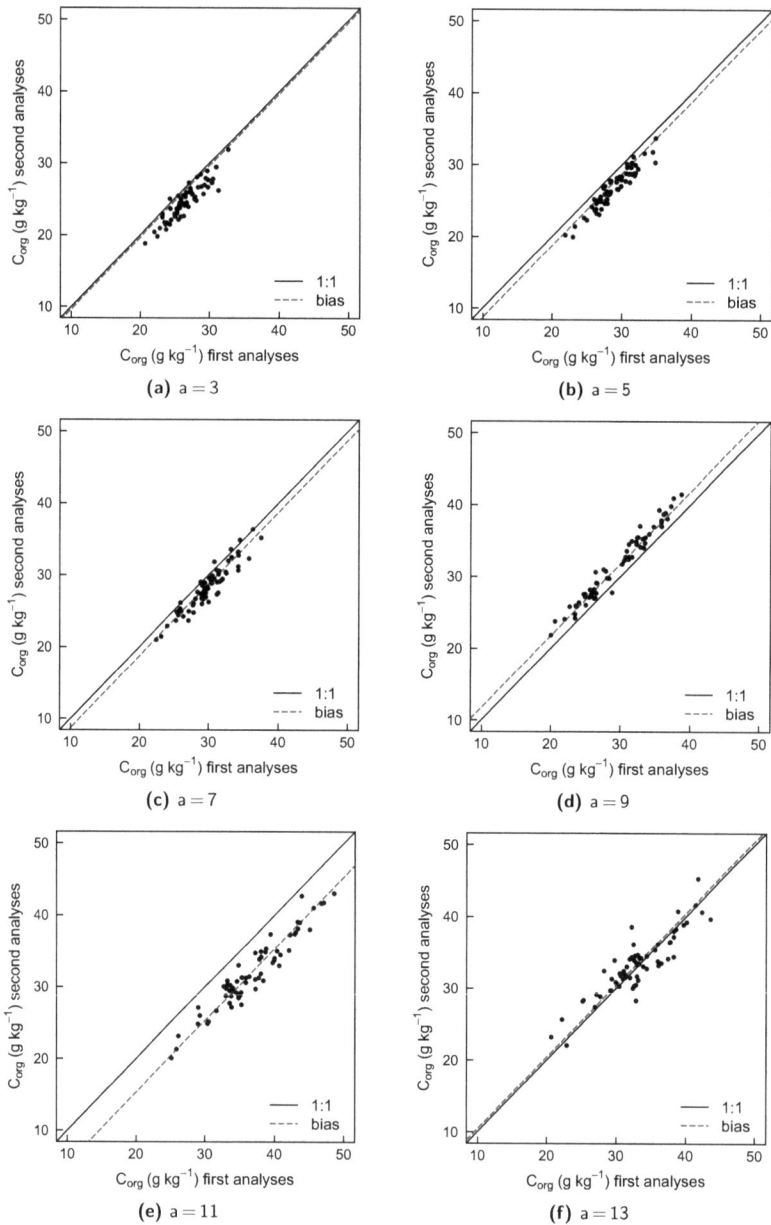

Figure A.14 First and second C_{org} analyses of *lu* samples by models derived from the *fr-so* dataset including a = 3,5,...,13 PLS factors.

A.2. Chapter 3: VNIRS laboratory models

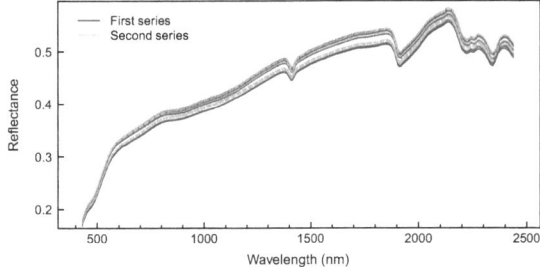

Figure A.15 Reflectance spectra of fine sand as recorded for different measurement series (the same subsample of sand and the same white reference panel were used for all measurements).

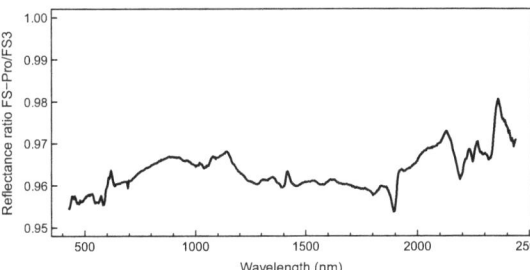

Figure A.16 Reflectance ratio of FS Pro to FS3 spectra of the sand sample used as standard. The ratio was calculated by comparing six spectra recorded by each spectrometer.

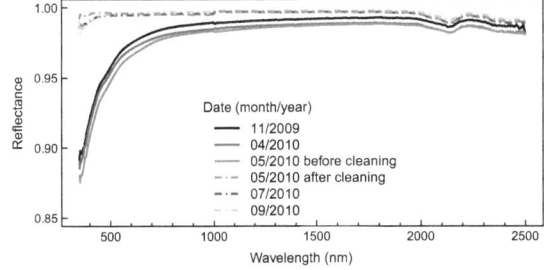

Figure A.17 Reflectance of the Spectralon panel used for white reference relative to an unused spectralon panel for different dates.

117

A. Additional information

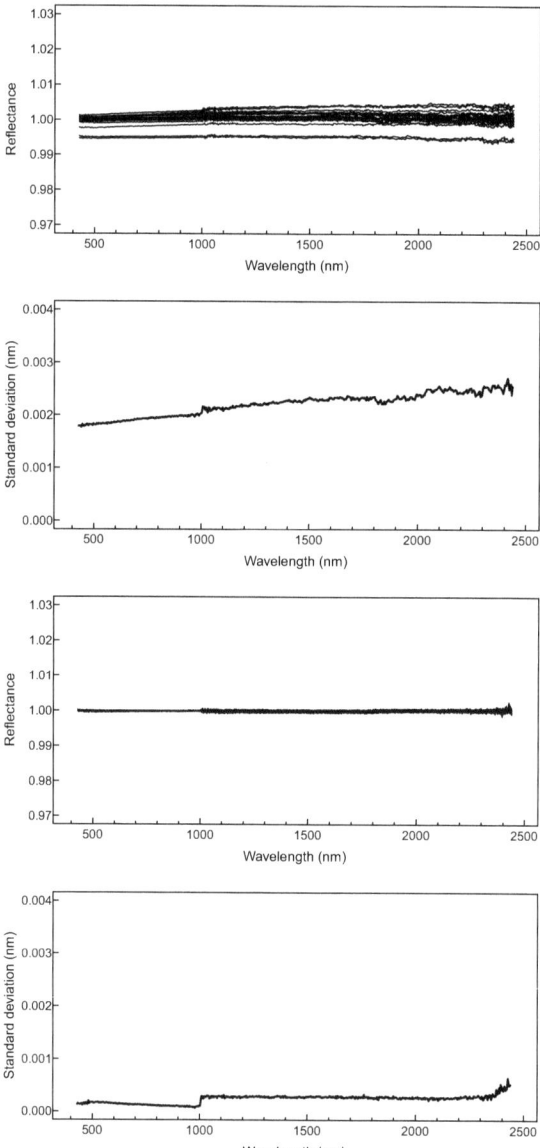

Figure A.18 Comparison of 20 measurements of a Spectralon panel with (top) and without dislocation (bottom) of the panel between measurements: reflectance spectra and their standard deviation.

A.3. Chapter 5: Soil moisture effects

Table A.1 Soil samples used for assessment of soil moisture effects on spectroscopic measurements. C_{tot} and N_{tot} determined by CN analyser, *pH* by Hellige pH indicator

Sample	Origin	Land use	pH	C_{tot} ($g\,kg^{-1}$)	N_{tot} ($g\,kg^{-1}$)
1	Wabern	crop rotation	5	36	4.3
2	Wabern	pasture	6	40	4.6
3	Wabern	pasture	6.5	26	2.5
4	Wabern	pasture	5.5	27	3.2
5	Kehrsatz	forest	4.5	77	5.0
6	Kehrsatz	crop rotation	5	28	3.1
7	Kehrsatz	crop rotation	5	26	2.9
8	Kehrsatz	pasture	7	37	3.8
9	Kehrsatz	crop rotation	7	57	4.2
10	Belp	fallow	7	40	1.5
11	Belp	crop rotation	7	55	3.8
12	Belp	crop rotation	7	66	5.0
13	Belp	pasture	7	43	4.6
14	Worb	pasture	7.5	71	4.8
15	Worb	forest	7.5	40	2.0
16	Kehrsatz	pasture	6.5	38	3.2
17	Niederbipp	pasture	4.5	45	5.1
18	Niederbipp	forest	5.5	55	4.0
19	Niederbipp	forest	4.0	105	7.5
20	Niederbipp	pasture	7	55	5.6
21	Niederbipp	crop rotation	7	53	3.9
22	Niederbipp	forest	4.0	49	3.2
23	Niederbipp	crop rotation	5	19	2.2
24	Niederbipp	crop rotation	6.5	22	2.3
25	Niederbipp	pasture	5.5	17	1.9
26	Lyss	pasture	6	69	5.7
27	Suberg	crop roation	5.5	22	2.7
28	Suberg	fallow	6.5	46	3.9
29	Lyss	crop rotation	7	11	1.4
30	Lyss	crop rotation	7	18	2.0
31	Radelfingen	crop rotation	4.5	13	1.4
32	Gurten	pasture	4.0	20	2.3
33	Gurten	forest	4.5	45	2.9
34	Kerzers	crop rotation	7	24	2.2
35	Kerzers	pasture	5.5	20	2.5
36	Kerzers	crop rotation	7	52	4.8
37	Kerzers	crop rotation	7	45	3.6
38	Ins	crop rotation	7	128	8.9
39	Ins	crop rotation	4.5	92	7.2
40	Brienz	crop rotation	6	28	3.1
41	Brienz	crop rotation	6.5	34	4.0
42	Galmiz	crop rotation	7	91	7.0
43	Wohlen	crop rotation	5	18	2.1

A. Additional information

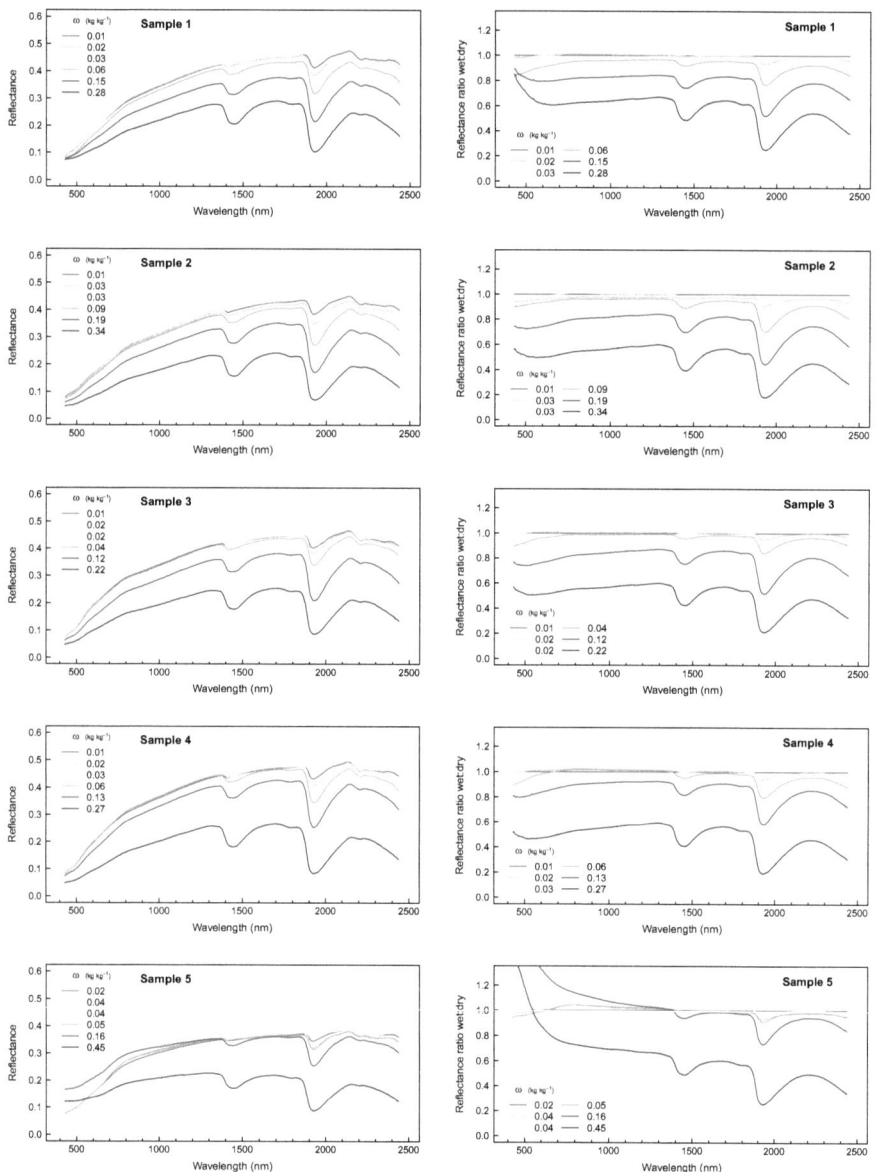

Figure A.19 Reflectance spectra (left) and reflectance ratio wet:dry (right) of all soil samples for different water contents. All data were corrected for packing errors.

A.3. Chapter 5: Soil moisture effects

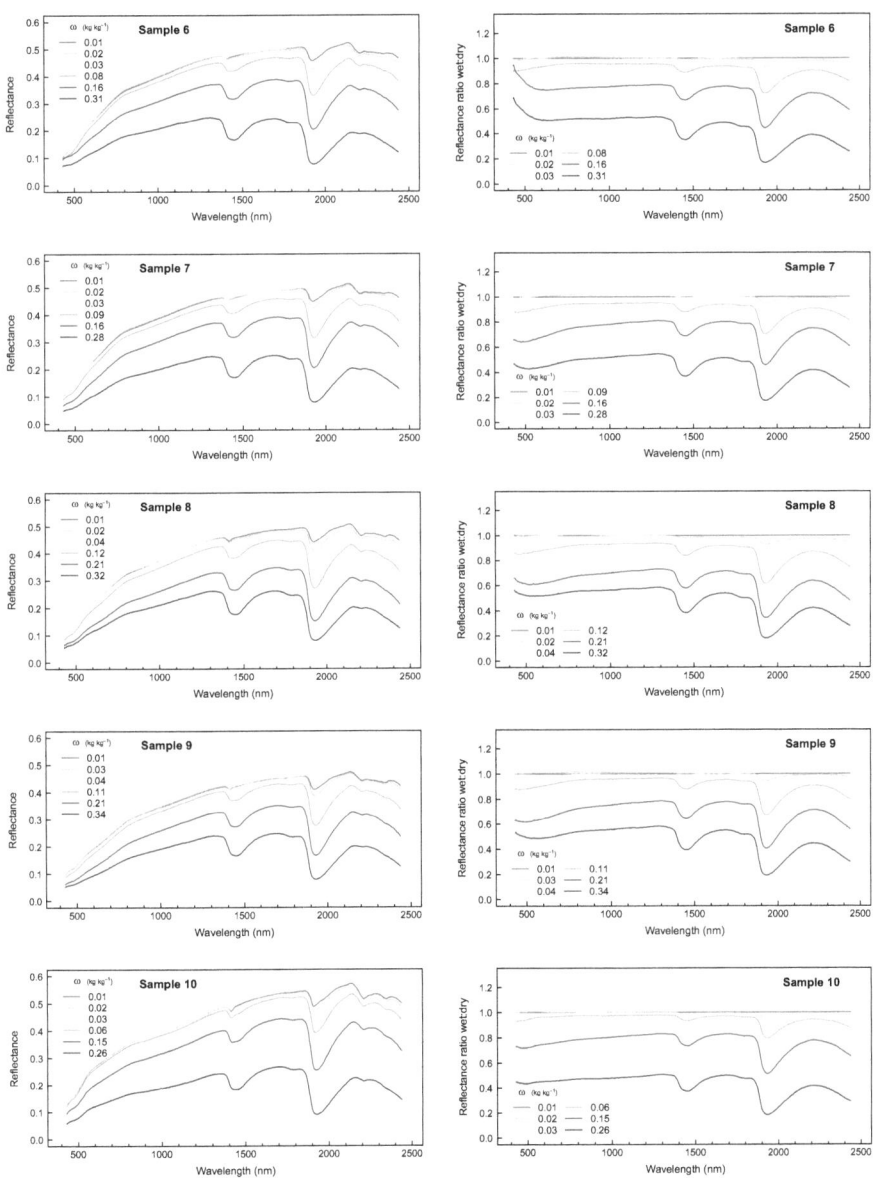

Figure A.19 (continued) Reflectance spectra (left) and reflectance ratio wet:dry (right) of all soil samples for different water contents. All data were corrected for packing errors.

A. Additional information

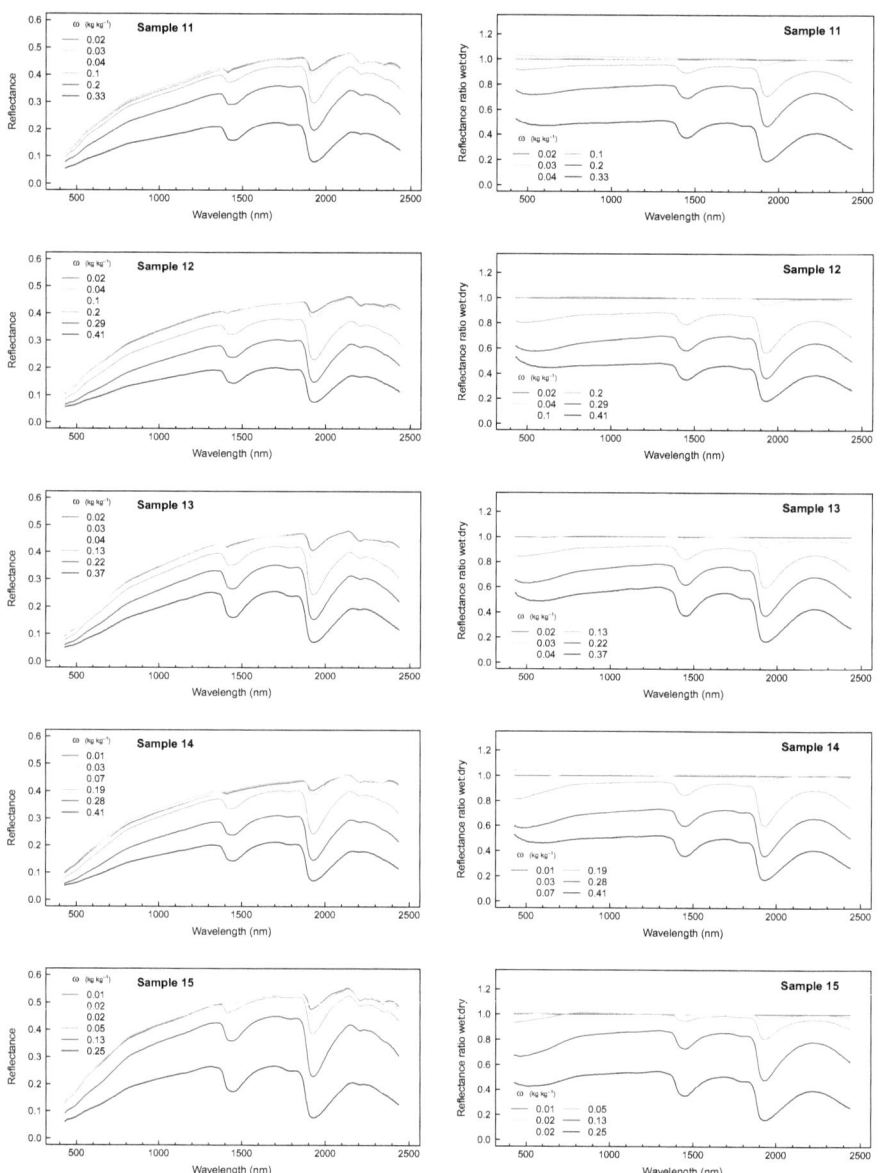

Figure A.19 (continued) Reflectance spectra (left) and reflectance ratio wet:dry (right) of all soil samples for different water contents. All data were corrected for packing errors.

A.3. Chapter 5: Soil moisture effects

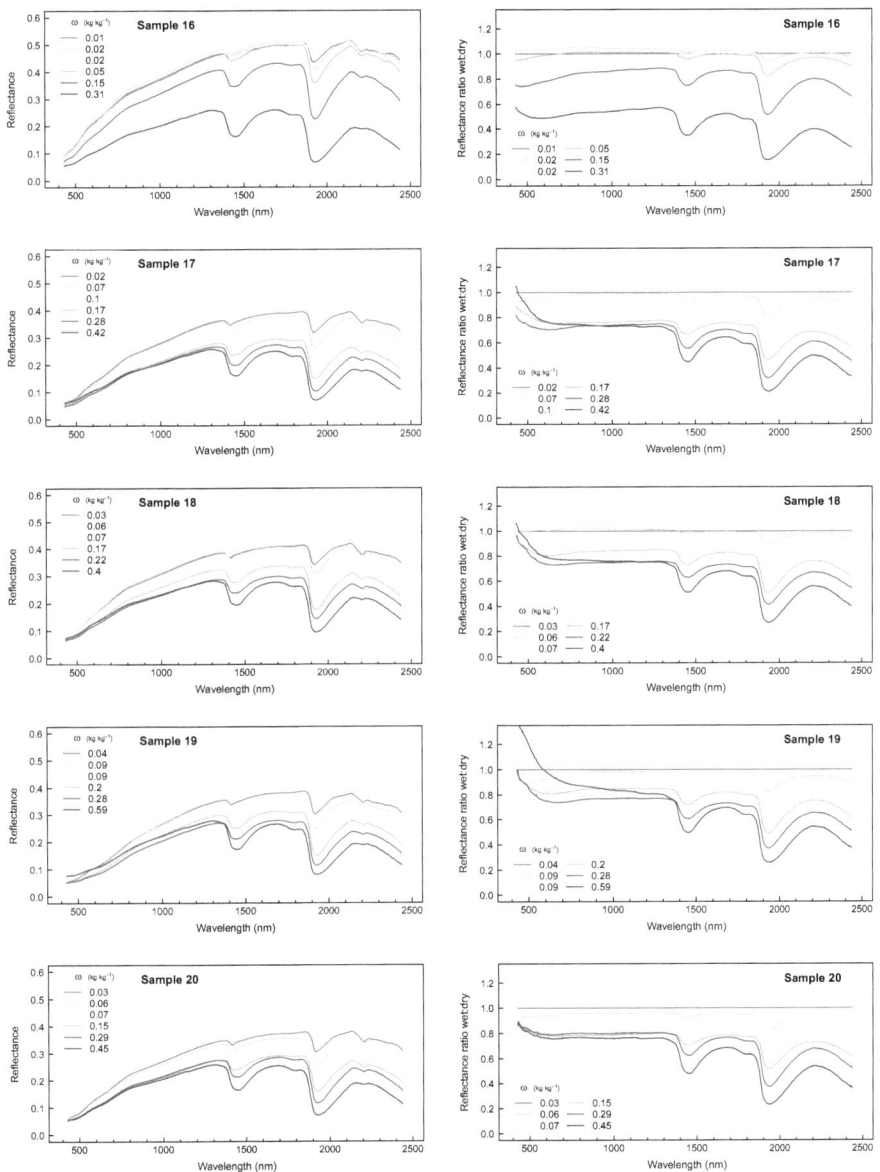

Figure A.19 (continued) Reflectance spectra (left) and reflectance ratio wet:dry (right) of all soil samples for different water contents. All data were corrected for packing errors.

123

A. Additional information

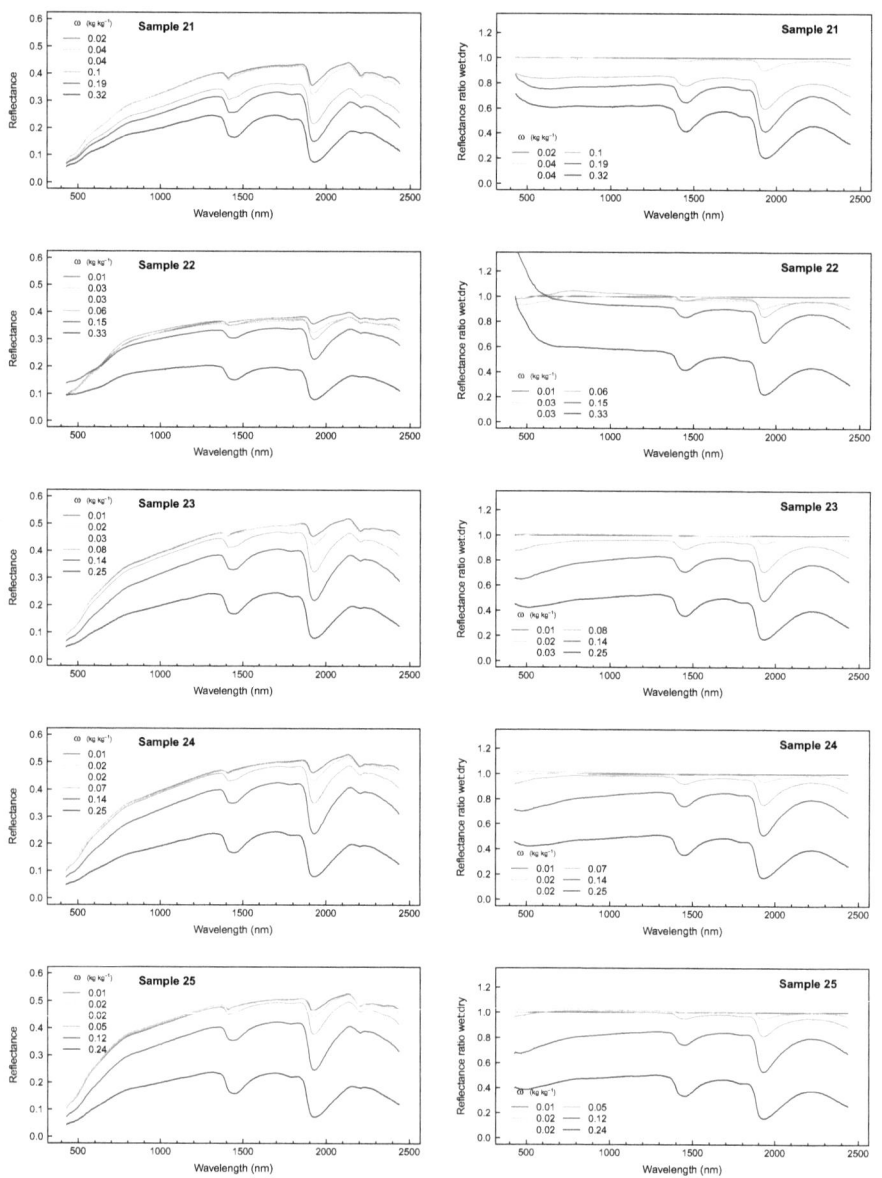

Figure A.19 (continued) Reflectance spectra (left) and reflectance ratio wet:dry (right) of all soil samples for different water contents. All data were corrected for packing errors.

A.3. Chapter 5: Soil moisture effects

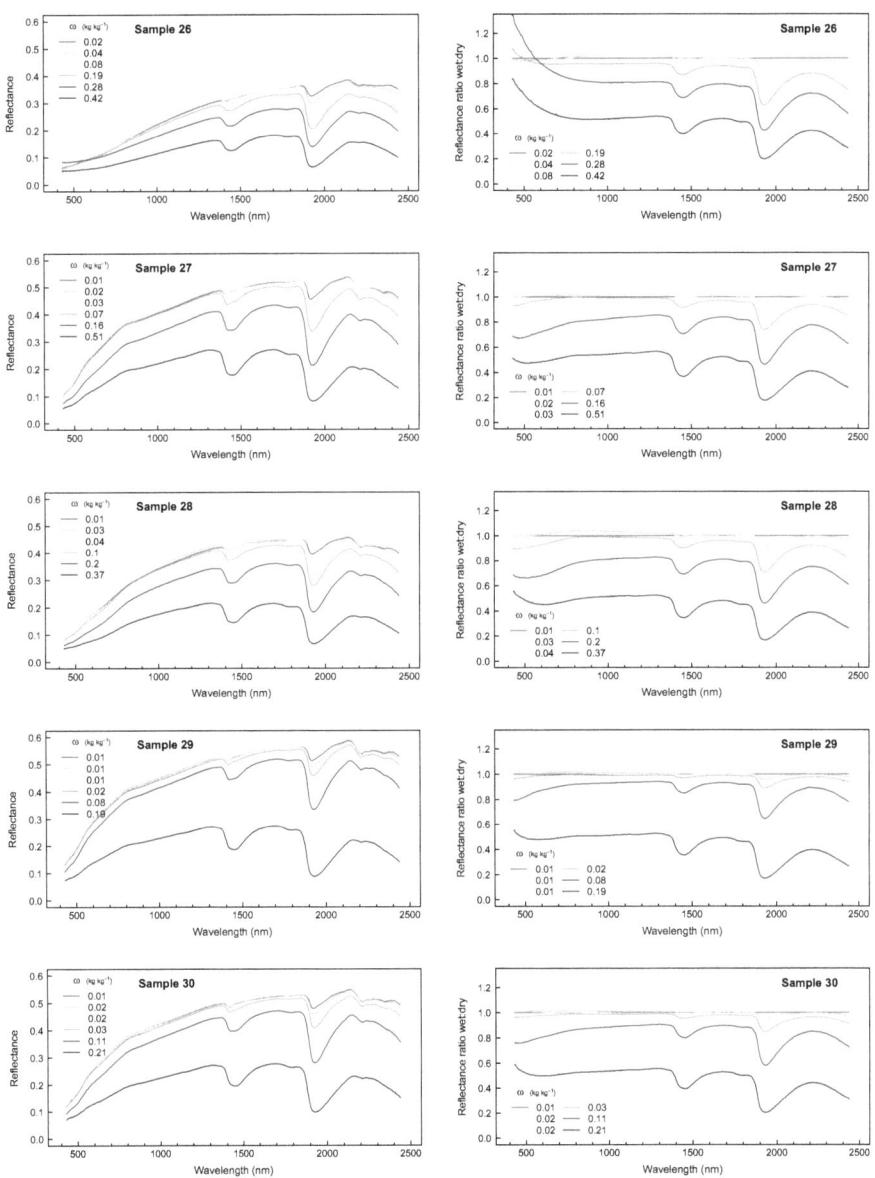

Figure A.19 (continued) Reflectance spectra (left) and reflectance ratio wet:dry (right) of all soil samples for different water contents. All data were corrected for packing errors.

A. Additional information

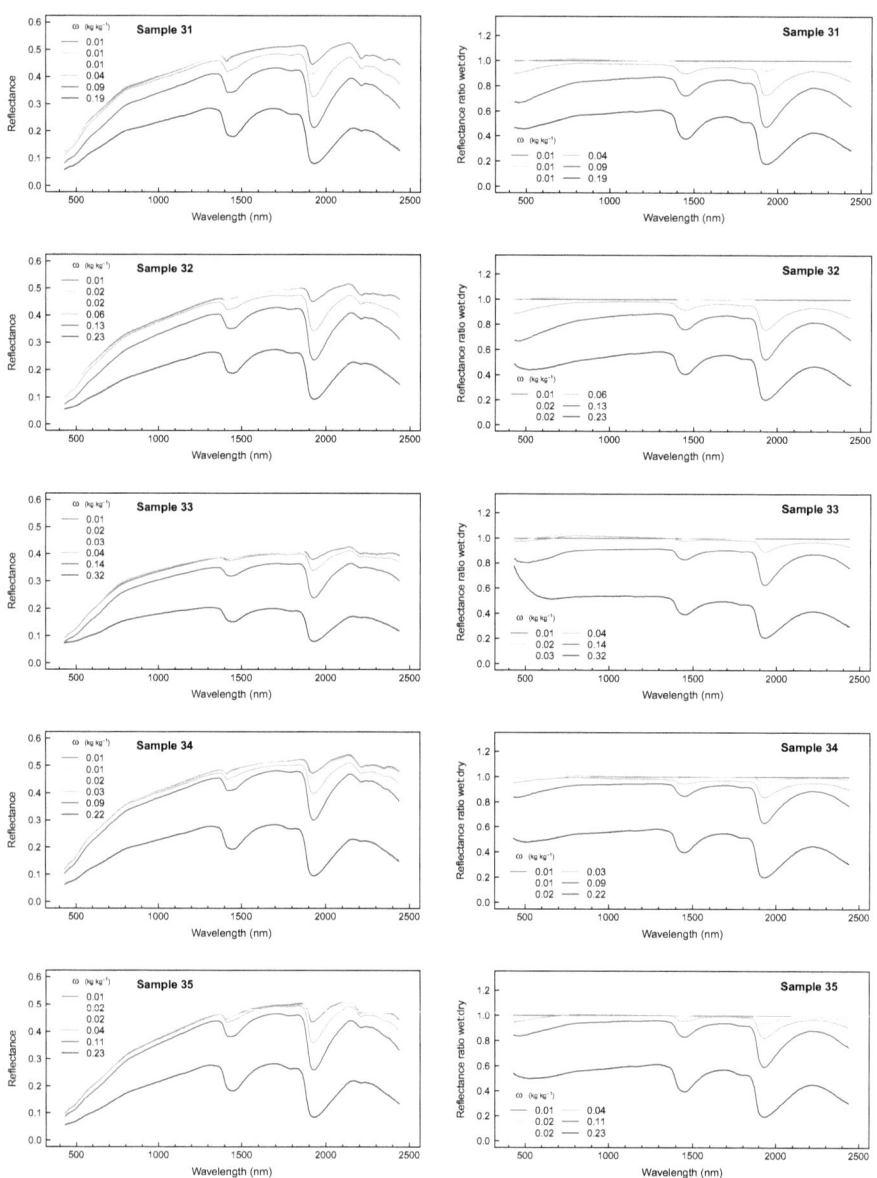

Figure A.19 (continued) Reflectance spectra (left) and reflectance ratio wet:dry (right) of all soil samples for different water contents. All data were corrected for packing errors.

A.3. Chapter 5: Soil moisture effects

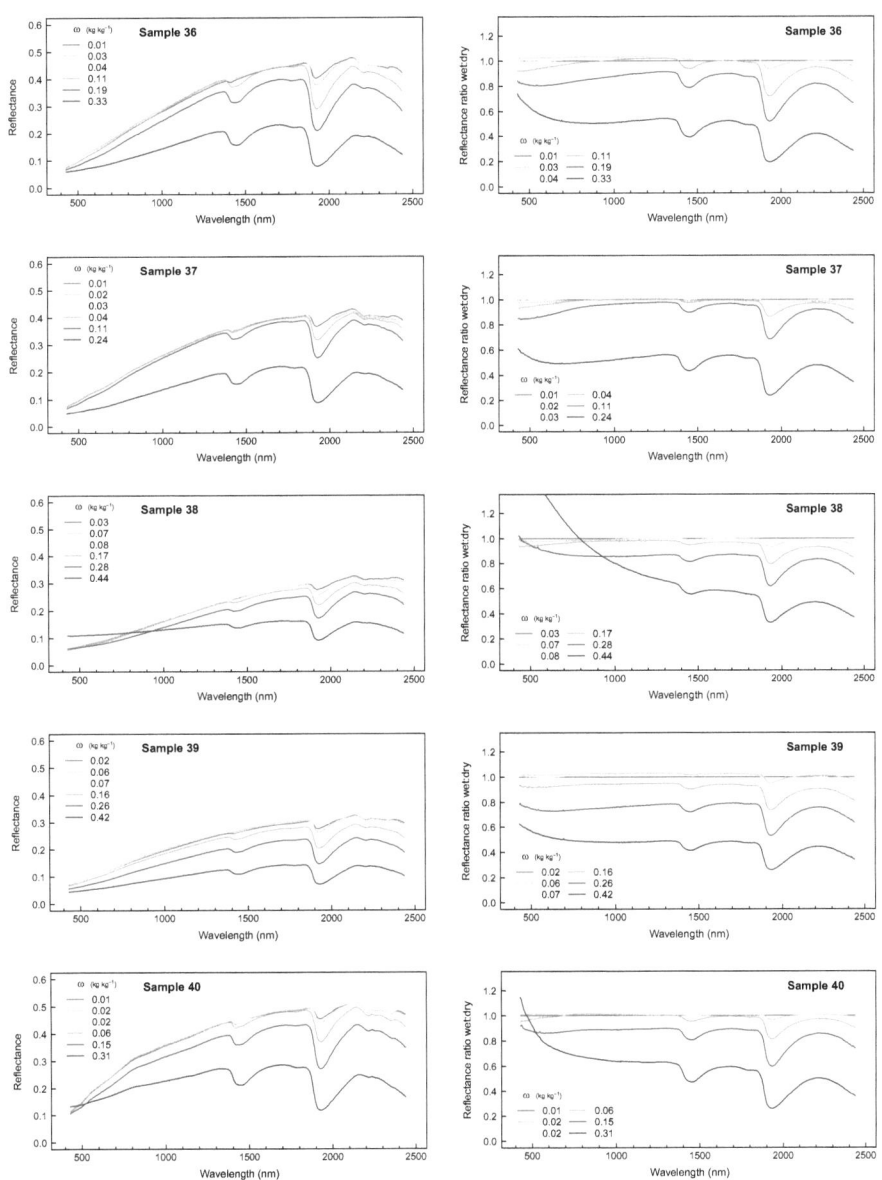

Figure A.19 (continued) Reflectance spectra (left) and reflectance ratio wet:dry (right) of all soil samples for different water contents. All data were corrected for packing errors.

A. Additional information

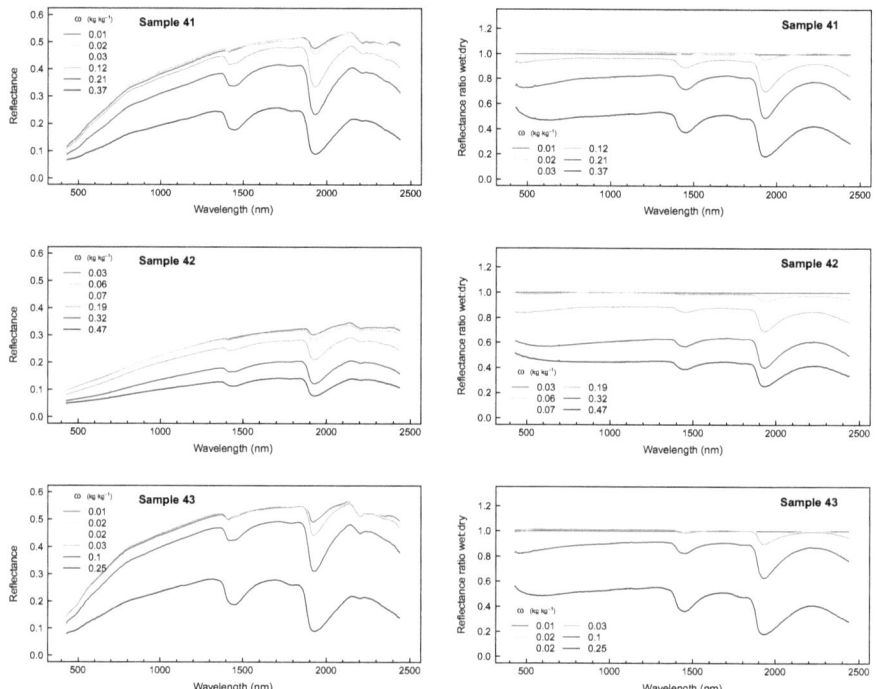

Figure A.19 (continued) Reflectance spectra (left) and reflectance ratio wet:dry (right) of all soil samples for different water contents. All data were corrected for packing errors.

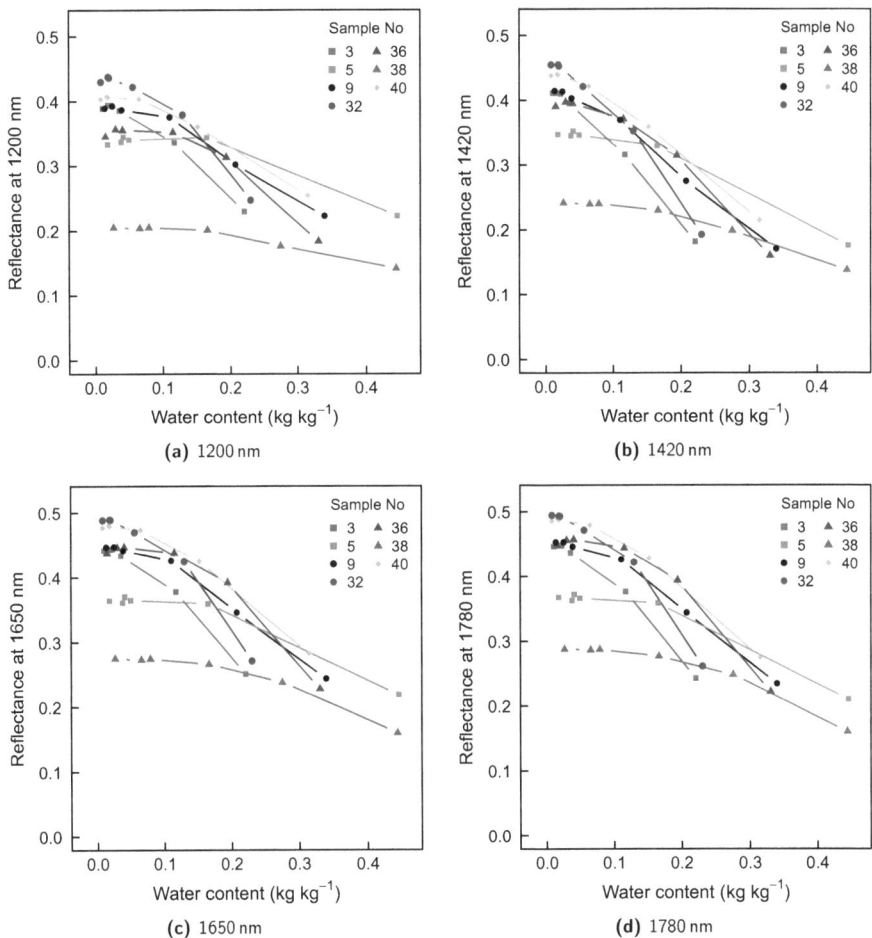

Figure A.20 Influence of soil moisture on reflectance at eight different wavelengths for seven randomly selected soil samples.

A. Additional information

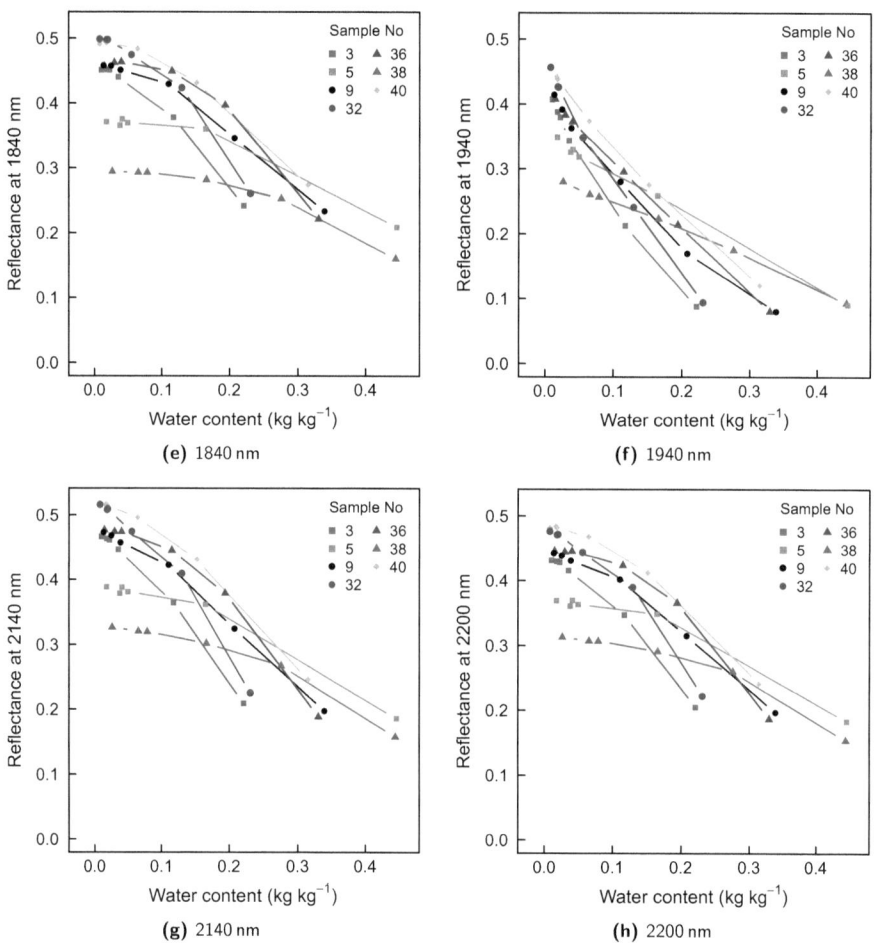

Figure A.20 (continued) Influence of soil moisture on reflectance at eight different wavelengths for seven randomly selected soil samples.

A.3. Chapter 5: Soil moisture effects

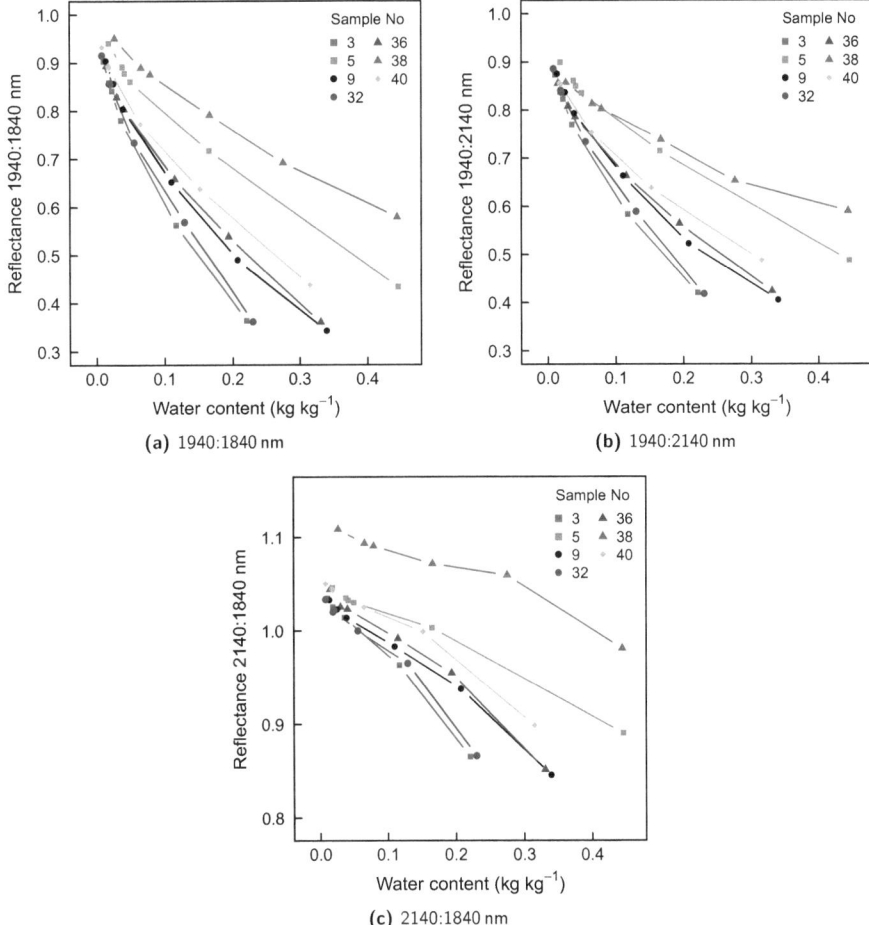

Figure A.21 Influence of soil moisture on reflectance ratios for seven randomly selected soil samples.

A. Additional information

A.3.1. Regression models

R output of the regression models to estimate water content ω (wc in R output) by F_1 or reflectance ratio 1940:1840 nm (rr in output), and the auxiliary variable C_{tot} (C.tot) presented in table 5.2 (page 69, section 5.3.2).

Linear model: $\omega \propto F_1$

```
Call:
lm(formula = wc ~ F1, data = d.a)

Residuals:
      Min        1Q    Median        3Q       Max
-0.162728 -0.021337 -0.006837  0.008981  0.270676

Coefficients:
             Estimate Std. Error t value Pr(>|t|)
(Intercept)   1.01185    0.02765   36.60   <2e-16 ***
F1           -1.04460    0.03193  -32.71   <2e-16 ***
---
Signif. codes:  0 '***' 0.001 '**' 0.01 '*' 0.05 '.' 0.1 ' ' 1

Residual standard error: 0.05396 on 256 degrees of freedom
Multiple R-squared: 0.8069,    Adjusted R-squared: 0.8062
F-statistic:  1070 on 1 and 256 DF,  p-value: < 2.2e-16
```

Quadratic model: $\omega \propto F_1 + (F_1)^2$

```
Call:
lm(formula = wc ~ F1 + I(F1^2), data = d.a)

Residuals:
      Min        1Q    Median        3Q       Max
-0.157574 -0.022156 -0.006632  0.010646  0.267456

Coefficients:
             Estimate Std. Error t value Pr(>|t|)
(Intercept)   0.8423     0.2418    3.484 0.000582 ***
F1           -0.6183     0.6045   -1.023 0.307363
I(F1^2)      -0.2625     0.3717   -0.706 0.480785
---
Signif. codes:  0 '***' 0.001 '**' 0.01 '*' 0.05 '.' 0.1 ' ' 1

Residual standard error: 0.05401 on 255 degrees of freedom
Multiple R-squared: 0.8073,    Adjusted R-squared: 0.8058
F-statistic: 534.2 on 2 and 255 DF,  p-value: < 2.2e-16
```

A.3. Chapter 5: Soil moisture effects

Quadratic model with auxiliary variable: $\omega \propto F_1 + (F_1)^2 + C_{tot} + C_{tot} \cdot (F_1)^2$

```
Call:
lm(formula = wc ~ C.tot + F1 + I(F1^2) + C.tot:I(F1^2),
    data = d.a)

Residuals:
     Min       1Q    Median       3Q       Max
-0.149220 -0.010644  0.000648  0.007379  0.236637

Coefficients:
                Estimate Std. Error t value Pr(>|t|)
(Intercept)     1.106852   0.153221   7.224 5.93e-12 ***
C.tot           0.063953   0.004221  15.152  < 2e-16 ***
F1             -1.971085   0.387394  -5.088 7.06e-07 ***
I(F1^2)         0.851790   0.242173   3.517 0.000517 ***
C.tot:I(F1^2)  -0.066566   0.005390 -12.351  < 2e-16 ***
---
Signif. codes:  0 '***' 0.001 '**' 0.01 '*' 0.05 '.' 0.1 ' ' 1

Residual standard error: 0.03406 on 253 degrees of freedom
Multiple R-squared: 0.924,    Adjusted R-squared: 0.9228
F-statistic:   769 on 4 and 253 DF,  p-value: < 2.2e-16
```

Linear model: $\omega \propto r_\lambda$

```
Call:
lm(formula = wc ~ rr, data = d.a)

Residuals:
    Min      1Q   Median      3Q      Max
-0.13735 -0.03171 -0.00637  0.01557  0.27512

Coefficients:
             Estimate Std. Error t value Pr(>|t|)
(Intercept)   0.50319    0.01295   38.85   <2e-16 ***
rr           -0.55295    0.01773  -31.19   <2e-16 ***
---
Signif. codes:  0 '***' 0.001 '**' 0.01 '*' 0.05 '.' 0.1 ' ' 1

Residual standard error: 0.05606 on 256 degrees of freedom
Multiple R-squared: 0.7917,    Adjusted R-squared: 0.7908
F-statistic: 972.7 on 1 and 256 DF,  p-value: < 2.2e-16
```

A. Additional information

Quadratic model: $\omega \propto r_\lambda + (r_\lambda)^2$

```
Call:
lm(formula = wc ~ rr + I(rr^2), data = d.a)

Residuals:
     Min        1Q    Median        3Q       Max
-0.155571 -0.023515 -0.007702  0.009651  0.282668

Coefficients:
            Estimate Std. Error t value Pr(>|t|)
(Intercept)  0.61430    0.04263  14.411  < 2e-16 ***
rr          -0.94266    0.14369  -6.560 2.97e-10 ***
I(rr^2)      0.30553    0.11181   2.732  0.00673 **
---
Signif. codes:  0 '***' 0.001 '**' 0.01 '*' 0.05 '.' 0.1 ' ' 1

Residual standard error: 0.05536 on 255 degrees of freedom
Multiple R-squared: 0.7976,     Adjusted R-squared: 0.796
F-statistic: 502.4 on 2 and 255 DF,  p-value: < 2.2e-16
```

Quadratic model with auxiliary variable: $\omega \propto r_\lambda + (r_\lambda)^2 + C_{tot} + C_{tot} \cdot r_\lambda$

```
Call:
lm(formula = wc ~ C.tot + rr + I(rr^2) + C.tot:rr, data = d.a)

Residuals:
     Min        1Q    Median        3Q       Max
-0.168544 -0.011881  0.000602  0.006729  0.236988

Coefficients:
             Estimate Std. Error t value Pr(>|t|)
(Intercept)  0.466453   0.030684  15.202  < 2e-16 ***
C.tot        0.051726   0.003716  13.921  < 2e-16 ***
rr          -0.983777   0.096254 -10.221  < 2e-16 ***
I(rr^2)      0.517607   0.075379   6.867 5.06e-11 ***
C.tot:rr    -0.053232   0.004958 -10.737  < 2e-16 ***
---
Signif. codes:  0 '***' 0.001 '**' 0.01 '*' 0.05 '.' 0.1 ' ' 1

Residual standard error: 0.03681 on 253 degrees of freedom
Multiple R-squared: 0.9112,     Adjusted R-squared: 0.9098
F-statistic: 649.3 on 4 and 253 DF,  p-value: < 2.2e-16
```

Bibliography

Alsberg, B. K., Woodward, A. M., Kell, D. B., 1997a. An introduction to wavelet transforms for chemometricians: A time-frequency approach. Chemometrics and Intelligent Laboratory Systems 37 (2), 215–239.

Alsberg, B. K., Woodward, A. M., Winson, M. K., Rowland, J., Kell, D. B., 1997b. Wavelet denoising of infrared spectra. Analyst 122 (7), 645–652.

ASD, 2009. Instrument specifications FieldSpec3 HR Full Range. Analytical Spectral Devices Inc., Boulder, CO, USA. http://www.asdi.com

ASD, 2011. Specifications High Intensity Muglight. Analytical Spectral Devices Inc., Boulder, CO, USA. http://www.asdi.com/accessories/high-intensity-muglight

Baumgardner, M. F., Silva, L. F., Biehl, L. L., Stoner, E. R., 1985. Reflectance properties of soils. Advances in Agronomy 38, 1–44.

Baxter, J., 2010. Ruée sur les terres africaines. Le Monde diplomatique 57 (670, Jan. 2010), 18.

Ben-Dor, E., Heller, D., Chudnovsky, A., 2008. A novel method of classifying soil profiles in the field using optical means. Soil Science Society of America Journal 72 (4), 1113–1123.

Ben-Gera, I., Norris, K. H., 1968. Direct spectrophotometric determination of fat and moisture in meat products. Journal of Food Science 33 (1), 64–67.

Bjørsvik, H. R., Martens, H., 2008. Data analysis: calibration of NIR instruments by PLS regression. In: Burns, D. A., Ciurczak, E. W. (eds.), Handbook of Near-Infrared Analysis. CRC Press, Boca Raton, Fla., pp. 189–205.

Bowers, S. A., Hanks, R. J., 1965. Reflection of radiant energy from soils. Soil Science 100 (2), 130–138.

Bremner, J. M., 1996. Nitrogen-total. In: Sparks, D. L. (ed.), Methods of Soil Analysis: Chemical Methods. SSSA Book Series 5. Soil Science Society of America, Madison, Wisc., pp. 1085–1122.

Brunet, D., Barthès, B. G., Chotte, J. L., Feller, C., 2007. Determination of carbon and nitrogen contents in Alfisols, Oxisols and Ultisols from Africa and Brazil using NIRS analysis: Effects of sample grinding and set heterogeneity. Geoderma 139 (1-2), 106–117.

Chang, C. W., Laird, D. A., Mausbach, M. J., Hurburgh, C. R., 2001. Near-infrared reflectance spectroscopy-principal components regression analyses of soil properties. Soil Science Society of America Journal 65 (2), 480–490.

Chang, G. W., Laird, D. A., Hurburgh, G. R., 2005. Influence of soil moisture on near-infrared reflectance spectroscopic measurement of soil properties. Soil Science 170 (4), 244–255.

Bibliography

Christy, C. D., 2008. Real-time measurement of soil attributes using on-the-go near infrared reflectance spectroscopy. Computers and Electronics in Agriculture 61 (1), 10–19.

Clark, R. N., 1999. Spectroscopy of rocks and minerals, and principles of spectroscopy. In: Rencz, A. N. (ed.), Manual of Remote Sensing, Volume 3, Remote Sensing for the Earth Sciences. John Wiley and Sons, New York, NY, pp. 3–58.

Cohen, A., Daubechies, I., Vial, P., 1993. Wavelets on the interval and fast wavelet transforms. Applied and Computational Harmonic Analysis 1 (1), 54–81.

Crawley, M. J., 2007. The R Book. Wiley, Chichester England , Hoboken N.J.

Daubechies, I., 2006. Ten Lectures on Wavelets, 9th edition. Vol. 61 of CBMS-NSF regional conference series in applied mathematics. Society for Industrial and Applied Mathematics, Philadelphia, Pa.

de Jong, S., 1993. SIMPLS - An alternative approach to partial least-squares regression. Chemometrics and Intelligent Laboratory Systems 18 (3), 251–263.

De Maesschalck, R., Jouan-Rimbaud, D., Massart, D. L., 2000. The Mahalanobis distance. Chemometrics and Intelligent Laboratory Systems 50 (1), 1–18.

Demattê, J. A. M., Campos, R. C., Alves, M. C., Fiorio, P. R., Nanni, M. R., 2004. Visible-NIR reflectance: a new approach on soil evaluation. Geoderma 121 (1-2), 95–112.

Demattê, J. A. M., Sousa, A. A., Alves, M. C., Nanni, M. R., Fiorio, P. R., Campos, R. C., 2006. Determining soil water status and other soil characteristics by spectral proximal sensing. Geoderma 135, 179–195.

FAL, 1996. Schweizerische Referenzmethoden der Eidg. Landwirtschaftlichen Forschungsanstalten. Zürich-Reckenholz.

FSO, 2001. The Changing Face of Land Use - Land Use Statistics of Switzerland. Federal Statistical Office, Neuchâtel.

FSO, 2011. Swiss Environmental Statistics - A Brief Guide 2011. Federal Statistical Office, Neuchâtel.

Fystro, G., 2002. The prediction of C and N content and their potential mineralisation in heterogeneous soil samples using Vis-NIR spectroscopy and comparative methods. Plant and Soil 246 (2), 139–149.

Gisi, U., Schenker, R., Schulin, R., Stadelmann, F. X., Sticher, H., 1997. Bodenökologie, 2nd edition. Thieme, Stuttgart.

Goldstein, D. H., Chenault, D. B., Pezzaniti, L., 2003. Polarimetric characterization of Spectralon. Air Force Research Laboratory, Munitions Directorate. Report AFRL-MN-EG-TR-2003-7013, pp. 16-26.

Gruijter, J., Brus, D., Bierkens, M., Knotters, M., 2006. Sampling for Natural Resource Monitoring. Springer, Berlin, Heidelberg.

Hastie, T., Tibshirani, R., Friedman, J., 2009. The Elements of Statistical Learning: Data Mining, Inference, and Prediction, 2nd edition. Springer Series in Statistics. Springer, New York, NY.

Hauert, C., 2007. Vergleich der Bodeneigenschaften im Direktsaat- und Pflugsystem mit Reflexionsspektroskopie und physikalischen Feldmethoden in den Gebieten Frienisberg und Oberaargau. Diploma thesis. CDE, University of Bern.

Hepperle, E., Lendi, M., 1993. Leben, Raum, Umwelt: Recht und Rechtspraxis. vdf, Zürich.

Hubert, M., Rousseeuw, P. J., van Aelst, S., 2008. High-breakdown robust multivariate methods. Statistical Science 23 (1), 92–119.

Hubert, M., Rousseeuw, P. J., Vanden Branden, K., 2005. ROBPCA: A new approach to robust principal component analysis. Technometrics 47 (1), 64–79.

Hubert, M., Vanden Branden, K., 2003. Robust methods for partial least squares regression. Journal of Chemometrics 17 (10), 537–549.

Hummel, J. W., Gaultney, L. D., Sudduth, K. A., 1996. Soil property sensing for site-specific crop management. Computers and Electronics in Agriculture 14 (2-3), 121–136.

Hunt, G. R., 1977. Spectral signatures of particulate minerals in the visible and near infrared. Geophysics 42 (3), 501–513.

Islam, K., McBratney, A., Singh, B., 2005. Rapid estimation of soil variability from the convex hull biplot area of topsoil ultra-violet, visible and near-infrared diffuse reflectance spectra. Geoderma 128 (3-4), 249–257.

iSOIL, 2008. Abstract iSOIL: Interactions between soil related sciences - Linking geophysics, soil science and digital soil mapping. http://www.isoil.info/

Lark, R. M., Webster, R., 1999. Analysis and elucidation of soil variation using wavelets. European Journal of Soil Science 50 (2), 185–206.

Leinweber, P., Schulten, H. R., Korschens, M., 1994. Seasonal variations of soil organic-matter in a long-term agricultural experiment. Plant and Soil 160 (2), 225–235.

Maleki, M. R., Mouazen, A. M., Ketelaere, B., Ramon, H., Baerdemaeker, J., 2008. On-the-go variable-rate phosphorus fertilisation based on a visible and near-infrared soil sensor. Biosystems Engineering 99, 35–46.

Maleki, M. R., Mouazen, A. M., Ramon, H., Baerdemaeker, J., 2007. Optimisation of soil VIS-NIR sensor-based variable rate application system of soil phosphorus. Soil and Tillage Research 94 (1), 239–250.

Mallat, S. G., 1989. A theory for multiresolution signal decomposition - the wavelet representation. IEEE Transaction on Pattern Analysis and Machine Intelligence 11 (7), 674–693.

McBratney, A. B., Minasny, B., Viscarra Rossel, R. A., 2006. Spectral soil analysis and inference systems: A powerful combination for solving the soil data crisis. Geoderma 136 (1-2), 272–278.

McBratney, A. B., Santos, M. L. M., Minasny, B., 2003. On digital soil mapping. Geoderma 117 (1-2), 3–52.

Möller, W., Nikolaus, K. P., Höpe, A., 2003. Degradation of the diffuse reflectance of Spectralon under low-level irradiation. Metrologia 40 (1), 212–215.

Mouazen, A. M., Baerdemaeker, J., Ramon, H., 2005a. Towards development of on-line soil moisture content sensor using a fibre-type NIR spectrophotometer. Soil and Tillage Research 80 (1-2), 171–183.

Mouazen, A. M., Karoui, R., Baerdemaeker, J., Ramon, H., 2005b. Classification of soil texture classes by using soil visual near infrared spectroscopy and factorial discriminant analysis techniques. Journal of Near Infrared Spectroscopy 13 (4), 231–240.

Mouazen, A. M., Karoui, R., Baerdemaeker, J., Ramon, H., 2006. Characterization of soil water content using measured visible and near infrared spectra. Soil Science Society of America Journal 70 (4), 1295–1302.

Mouazen, A. M., Karoui, R., Deckers, J., Baerdemaeker, J., Ramon, H., 2007a. Potential of visible and near-infrared spectroscopy to derive colour groups utilising the Munsell soil colour charts. Biosystems Engineering 97 (2), 131–143.

Mouazen, A. M., Maleki, M. R., Baerdemaeker, J., Ramon, H., 2007b. On-line measurement of some selected soil properties using a VIS-NIR sensor. Soil and Tillage Research 93 (1), 13–27.

Naes, T., 2002. A User-Friendly Guide to Multivariate Calibration and Classification. NIR Publications, Chichester.

Nason, G. P., 2008. Wavelet Methods in Statistics with R. Springer, New York, NY.

Nelson, D. W., Sommers, L. E., 1996. Total carbon, organic carbon and organic matter. In: Sparks, D. L. (ed.), Methods of Soil Analysis: Chemical Methods. SSSA Book Series 5. Soil Science Society of America, Madison, Wisc., pp. 961–1069.

Odlare, M., Svensson, K., Pell, M., 2005. Near infrared reflectance spectroscopy for assessment of spatial soil variation in an agricultural field. Geoderma 126 (3-4), 193–202.

Pimstein, A., Notesco, G., Ben-Dor, E., 2011. Performance of three identical spectrometers in retrieving soil reflectance under laboratory conditions. Soil Science Society of America Journal 75 (2), 746–759.

R Development Core Team, 2011. R: A Language and Environment for Statistical Computing. http://www.R-project.org

Read, D., Beerling, D., Cannell, M., Cox, P., Curran, P., Grace, J., Ineson, P., Jarvis, P., Malhi, Y., Powlson, D., Shepherd, J., I.Woodward, I., 2001. The Role of Land Carbon Sinks in Mitigating Global Climate Change. Royal Society, London.

Ruth, L., 2010. Bestimmung des Humusgehalts von landwirtschaftlich genutzten Böden im Oberaargau mit Hilfe der Reflexionsspektroskopie. Diploma thesis. CDE, University of Bern.

Savitzky, A., Golay, M. J., 1964. Smoothing and differentiation of data by simplified least squares procedures. Analytical Chemistry 36 (8), 1627.

Schumacher, B. A., 2002. Methods for the determination of total organic carbon (TOC) in soils and sediments. Ecological Risk Assessment Support Center, Office of Research and Development, U.S. Environmental Protection Agency. Las Vegas.

Seiler, B., 2006. Quantitative assessment of soil parameters in Western Tajikistan. Diploma thesis. Department of Geography, University of Zurich.

Siesler, H. W., 2008. Basic principles of near-infrared spectroscopy. In: Burns, D. A., Ciurczak, E. W. (eds.), Handbook of Near-Infrared Analysis. CRC Press, Boca Raton, Fla.

Skidmore, E. L., Dickerson, J. D., Schimmelpfennig, H., 1975. Evaluating surface-soil water-content by measuring reflectance. Soil Science Society of America Journal 39 (2), 238–242.

Stenberg, B., 2010. Effects of soil sample pretreatments and standardised rewetting as interacted with sand classes on Vis-NIR predictions of clay and soil organic carbon. Geoderma 158 (1-2), 15–22.

Stenberg, B., Viscarra Rossel, R. A., Mouazen, A. M., Wetterlind, J., 2010. Visible and near infrared spectroscopy in soil science. Advances in Agronomy 107, 163–215.

Stevens, A., van Wesemael, B., Bartholomeus, H., Rosillon, D., Tychon, B., Ben-Dor, E., 2008. Laboratory, field and airborne spectroscopy for monitoring organic carbon content in agricultural soils. Geoderma 144 (1-2), 395–404.

Stevens, A., van Wesemael, B., Vandenschrick, G., Toure, S., Tychon, B., 2006. Detection of carbon stock change in agricultural soils using spectroscopic techniques. Soil Science Society of America Journal 70 (3), 844–850.

Sudduth, K. A., Hummel, J. W., 1993a. Portable, near-infrared spectrophotometer for rapid soil analysis. Transactions of the ASAE 36 (1), 185–193.

Sudduth, K. A., Hummel, J. W., 1993b. Soil organic-matter, CEC, and moisture sensing with a portable NIR spectrophotometer. Transactions of the ASAE 36 (6), 1571–1582.

Udelhoven, T., Emmerling, C., Jarmer, T., 2003. Quantitative analysis of soil chemical properties with diffuse reflectance spectrometry and partial least-square regression: A feasibility study. Plant and Soil 251 (2), 319–329.

Viscarra Rossel, R. A., 2009. The Soil Spectroscopy Group and the development of a global soil spectral library. NIR news 20 (4), 14–15.

Viscarra Rossel, R. A., Behrens, T., 2010. Using data mining to model and interpret soil diffuse reflectance spectra. Geoderma 158 (1-2), 46–54.

Viscarra Rossel, R. A., Cattle, S. R., Ortega, A., Fouad, Y., 2009. In situ measurements of soil colour, mineral composition and clay content by vis-NIR spectroscopy. Geoderma 150 (3-4), 253–266.

Viscarra Rossel, R. A., Jeon, Y. S., Odeh, I. O. A., McBratney, A. B., 2008. Using a legacy soil sample to develop a mid-IR spectral library. Australian Journal of Soil Research 46 (1), 1–16.

Viscarra Rossel, R. A., Lark, R. M., 2009. Improved analysis and modelling of soil diffuse reflectance spectra using wavelets. European Journal of Soil Science 60 (3), 453–464.

Viscarra Rossel, R. A., McGlynn, R. N., McBratney, A. B., 2006a. Determining the composition of mineral-organic mixes using UV-vis-NIR diffuse reflectance spectroscopy. Geoderma 137 (1-2), 70–82.

Viscarra Rossel, R. A., Walvoort, D. J. J., McBratney, A. B., Janik, L. J., Skjemstad, J. O., 2006b. Visible, near infrared, mid infrared or combined diffuse reflectance spectroscopy for simultaneous assessment of various soil properties. Geoderma 131 (1-2), 59–75.

Voss, K. J., Zhang, H., 2006. Bidirectional reflectance of dry and submerged Labsphere Spectralon plaque. Applied Optics 45 (30), 7924–7927.

Waiser, T. H., Morgan, C. L. S., Brown, D. J., Hallmark, C. T., 2007. In situ characterization of soil clay content with visible near-infrared diffuse reflectance spectroscopy. Soil Science Society of America Journal 71 (2), 389–396.

Walkley, A., Black, I. A., 1934. An examination of the Degtjareff method for determining soil organic matter, and a proposed modification of the chromic acid titration method. Soil Science 37 (1), 29–38.

Weidner, V. R., Hsia, J. J., 1981. Reflection properties of pressed polytetrafluoroethylene powder. Journal of the Optical Society of America 71 (7), 856–861.

Whiting, M. L., Li, L., Ustin, S. L., 2004. Predicting water content using Gaussian model on soil spectra. Remote Sensing of Environment 89 (4), 535–552.

Wold, S., Sjostrom, M., Eriksson, L., 2001. PLS-regression: a basic tool of chemometrics. Chemometrics and Intelligent Laboratory Systems 58 (2), 109–130.

Workman, J., Shenk, J., 2004. Understanding and using the near-infrared spectrum as an analytical method. In: Roberts, C. A., Workman, J., Reeves, J. B. (eds.), Near-Infrared Spectroscopy in Agriculture. Vol. no. 44 of Agronomy. American Society of Agronomy, Madison, Wis, pp. 3–10.

Workman, J., Weyer, L., 2007. Practical Guide to Interpretive Near-Infrared Spectroscopy. CRC Press, Boca Raton, Fla.

Workman, J. J., 2008. NIR spectroscopy calibration basics. In: Burns, D. A., Ciurczak, E. W. (eds.), Handbook of Near-Infrared Analysis. CRC Press, Boca Raton, Fla.

i want morebooks!

Buy your books fast and straightforward online - at one of world's fastest growing online book stores! Environmentally sound due to Print-on-Demand technologies.

Buy your books online at
www.get-morebooks.com

Kaufen Sie Ihre Bücher schnell und unkompliziert online – auf einer der am schnellsten wachsenden Buchhandelsplattformen weltweit! Dank Print-On-Demand umwelt- und ressourcenschonend produziert.

Bücher schneller online kaufen
www.morebooks.de

VDM Verlagsservicegesellschaft mbH
Heinrich-Böcking-Str. 6-8   Telefon: +49 681 3720 174   info@vdm-vsg.de
D - 66121 Saarbrücken   Telefax: +49 681 3720 1749 www.vdm-vsg.de

Printed by Books on Demand GmbH, Norderstedt / Germany